Institut für Baustatik und Konstruktion, ETH Zürich

W0051050

Strength and Deformations of Structural Concrete Subjected to In-Plane Shear and Normal Forces

Walter Kaufmann

Springer Basel AG

IBK Bericht Nr. 234, Juli

KEYWORDS: bond, compression field approach, compression softening, limit analysis, plane stress state, reinforced concrete, tension stiffening

Dieses Werk ist urheberrechtlich geschützt. Die dadurch begründeten Rechte, insbesondere die der Uebersetzung, des Nachdrucks, des Vortrags, der Entnahme von Abbildungen und Tabellen, der Funksendung, der Mikroverfilmung oder der Vervielfältigung auf anderen Wegen und der Speicherung in Datenverarbeitungsanlagen, bleiben, auch bei nur auszugsweiser Verwertung, vorbehalten. Eine Vervielfältigung dieses Werkes oder von Teilen dieses Werkes ist auch im Einzelfall nur in den Grenzen der gesetzlichen Bestimmungen des Urheberrechtsgesetzes in der jeweils geltenden Fassung zulässig. Sie ist grundsätzlich vergütungspflichtig. Zuwiderhandlungen unterliegen den Strafbestimmungen des Urheberrechts.

© 1998 Springer Basel AG
Originally published by Birkhäuser Verlag Basel, in 1998
Gedruckt auf säurefreiem Papier

ISBN 978-3-7643-5989-8   ISBN 978-3-0348-7612-4 (eBook)
DOI 10.1007/978-3-0348-7612-4

987654321

Strength and Deformations of Structural Concrete Subjected to In-Plane Shear and Normal Forces

Walter Kaufmann

Institute of Structural Engineering
Swiss Federal Institute of Technology Zurich

Zurich
July 1998

Preface

The present doctoral thesis was developed within the framework of the research project "Deformation Capacity of Structural Concrete". This project aims at developing a consistent and experimentally verified theory of the deformation capacity of structural concrete. Previous work included the development of a theoretical model, the so-called Tension Chord Model, which allows a comprehensive description of the load-deformation behaviour of tension members in non-prestressed and prestressed concrete structures.

The present work focuses on a new theoretical model, the so-called Cracked Membrane Model. For members subjected to in-plane forces this new model combines the basic concepts of the modified compression field theory and the tension chord model. Crack spacings and tension stiffening effects in cracked membranes are determined from first principles and the link to plasticity theory methods is maintained since equilibrium conditions are formulated in terms of stresses at the cracks rather than average stresses between the cracks.

The research project "Deformation Capacity of Structural Concrete" has been funded by the Swiss National Science Foundation and the Association of the Swiss Cement Producers. This support is gratefully acknowledged.

Zurich, July 1998   Prof. Dr. Peter Marti

Abstract

This thesis aims at contributing to a better understanding of the load-carrying and deformational behaviour of structural concrete subjected to in-plane shear and normal forces. Simple, consistent physical models reflecting the influences of the governing parameters are developed on whose basis (i) a realistic assessment of the deformation capacity of structural concrete subjected to in-plane loading is possible, (ii) the limits of applicability of the theory of plasticity to structural concrete can be explored, and (iii) current design provisions can be critically reviewed, supplemented and harmonised.

In the first part of this thesis relevant properties of concrete and reinforcement are examined, basic aspects of the theory of plasticity and its application to structural concrete are summarised, previous work on plane stress in structural concrete is reviewed, and fundamental aspects of the behaviour of cracked concrete membranes are investigated.

In the second part a new model for cracked, orthogonally reinforced concrete panels subjected to a homogeneous state of plane stress is presented. The cracked membrane model combines the basic concepts of original compression field approaches and a recently developed tension chord model. Crack spacings and tensile stresses between the cracks are determined from first principles and the link to limit analysis methods is maintained since equilibrium conditions are expressed in terms of stresses at the cracks rather than average stresses between the cracks. Both a general numerical method and an approximate analytical solution are derived and the results are compared with previous theoretical and experimental work. Simple expressions for the ultimate load of reinforced concrete panels in terms of the reinforcement ratios and the cylinder compressive strength of concrete are proposed, the influences of prestressing and axial forces are examined and basic aspects of the behaviour of uniaxially reinforced panels are discussed.

In the third part the behaviour of beams in shear is examined, focusing on simplified models for girders with flanged cross-section. For regions where all static and geometric quantities vary only gradually along the girder axis a procedure is presented that allows carrying out load-deformation analyses of the web, accounting for tension stiffening of the stirrups and the variation of the principal compressive stress direction over the depth of the cross-section. The results are compared with typical design assumptions and with previous work, justifying the usual design assumption of a uniform uniaxial compressive stress field in the web. Discontinuity regions characterised by abrupt changes of static quantities are analysed using the practically relevant case of the support region of a constant-depth girder with flanged cross-section as an illustrative example. Fan-shaped discontinuous stress fields with variable concrete compressive strength are examined and a method that allows checking whether the concrete stresses are below the concrete compressive strength throughout the fan region is presented. The results are compared with typical design assumptions and with previous experimental work. A previously suggested design procedure for support regions is supplemented and justified.

The fourth part summarises and discusses the results obtained in the first three parts of this thesis and concludes with a set of recommendations for future research.

Kurzfassung

Diese Dissertation soll zu einem besseren Verständnis des Trag- und Verformungsverhaltens von Stahlbeton unter ebener Beanspruchung beitragen. Es werden einfache und konsistente physikalische Modelle entwickelt, welche die massgebenden Einflüsse erfassen und auf deren Basis (i) eine realistische Beurteilung des Verformungsvermögens von Stahlbeton im ebenen Spannungszustand möglich ist, (ii) die Grenzen der Anwendbarkeit der Plastizitätstheorie auf Stahlbeton erforscht und (iii) gängige Bemessungsvorschriften kritisch beurteilt, ergänzt und harmonisiert werden können.

Im ersten Teil werden relevante Eigenschaften von Beton und Bewehrung untersucht, Grundzüge der Plastizitätstheorie und ihrer Anwendung auf Stahlbeton zusammengefasst, frühere Arbeiten über Stahlbeton im ebenen Spannungszustand erörtert und das Verhalten von gerissenen Betonscheibenelementen analysiert.

Im zweiten Teil wird ein neues Modell für gerissene, orthogonal bewehrte Betonscheibenelemente unter homogener ebener Beanspruchung vorgelegt. Das Gerissene Scheibenmodell kombiniert die Grundkonzepte der Druckfeldtheorie und eines unlängst entwickelten Zuggurtmodells. Rissabstände und Zugspannungen zwischen den Rissen werden von mechanischen Grundprinzipien abgeleitet, und die Verbindung zu Traglastverfahren bleibt erhalten, da Gleichgewicht in Spannungen an den Rissen – und nicht in mittleren Spannungen zwischen den Rissen – formuliert wird. Ein allgemeines numerisches Verfahren sowie eine analytische Näherungslösung werden hergeleitet und die Resultate mit früheren theoretischen und experimentellen Arbeiten verglichen. Einfache Ausdrücke für die Traglast in Funktion der Bewehrungsgehalte und der Zylinderdruckfestigkeit des Betons werden angegeben, der Einfluss von Vorspannung und Normalkräften wird untersucht und Grundzüge des Verhaltens von einachsig bewehrten Elementen werden diskutiert.

Im dritten Teil wird das Verhalten schubbeanspruchter profilierter Träger anhand vereinfachter Modelle untersucht. Für Bereiche, in welchen alle statischen und geometrischen Grössen entlang der Trägerachse nur allmählich variieren, wird ein Verfahren hergeleitet, welches Last-Verformungsanalysen des Steges unter Berücksichtigung des Verbundes der Bügel und der Variation der Hauptdruckspannungsrichtung über die Querschnittshöhe ermöglicht. Die Resultate werden mit gängigen Bemessungsannahmen und früheren Arbeiten verglichen, und es wird gezeigt, dass die übliche Annahme eines homogenen einachsigen Druckspannungsfeldes im Steg gerechtfertigt ist. Diskontinuitätsbereiche mit sprunghaft veränderlichen statischen Grössen werden anhand des Auflagerbereichs eines parallelgurtigen profilierten Trägers untersucht. Fächerförmige diskontinuierliche Spannungsfelder mit veränderlicher Betondruckfestigkeit werden analysiert, und ein Verfahren wird hergeleitet, mit welchem überprüft werden kann, ob die Spannungen im gesamten Fächerbereich unterhalb der Betondruckfestigkeit liegen. Die Resultate werden mit gängigen Bemessungsannahmen und früheren Arbeiten verglichen. Ein früher vorgeschlagenes Bemessungsverfahren für Auflagerbereiche wird ergänzt und seine Berechtigung wird nachgewiesen.

Im vierten Teil werden die Resultate zusammengefasst und diskutiert sowie eine Reihe von Möglichkeiten für weiterführende Forschungsarbeiten aufgezeigt.

Résumé

Cette thèse vise à contribuer à une meilleure compréhension de la capacité portante et du mode de déformation du béton armé soumis à des sollicitations membranaires par efforts normaux et tranchants. Des modèles simples, consistants et tenants compte des paramètres principaux sont développés, permettant (i) d'évaluer de façon réaliste la capacité de déformation des structures en béton armé soumises à des sollicitations membranaires, (ii) d'explorer les limites de l'application de la théorie de la plasticité au béton armé, et (iii) de réviser, compléter et harmoniser les règles de dimensionnement contemporaines.

La première partie examine les propriétés essentielles du béton et de l'acier, récapitule les bases de la théorie de la plasticité et de son application au béton armé, discute les études antérieures sur le sujet du béton armé soumis à des sollicitations membranaires et étudie les aspects fondamentaux du comportement des membranes fissurées en béton.

La seconde partie présente un nouveau modèle pour les panneaux fissurés en béton à armature orthogonale soumis à des sollicitations membranaires homogènes. Le modèle de la membrane fissurée combine les concepts de base des modèles de champ de compression et d'un modèle de membrure en traction récemment développé. Les espacements des fissures et les contraintes entre les fissures sont déterminés à partir des principes mécaniques de base et le lien aux concepts de l'analyse limite est maintenu en exprimant l'équilibre en termes des contraintes aux fissures, plutôt qu'en termes des contraintes moyennes entre les fissures. Une méthode numérique générale et une solution analytique approximative sont développées et les résultats sont comparés aux études théoriques et expérimentales précédentes. Des expressions simples pour la charge limite en fonction des taux d'armature et de la résistance du béton à la compression sur cylindre sont proposées, les influences de la précontrainte et de forces normales sont examinées et le comportement des panneaux armés unidirectionellement est examiné.

La troisième partie examine le comportement des poutres profilées dont seule l'âme résiste à l'effort tranchant, en utilisant des modèles simplifiés. Pour les régions où les grandeurs statiques et géométriques ne changent que graduellement le long de l'axe de la poutre une méthode permettant d'effectuer des analyses de charge-déformation de l'âme est présentée, tenant compte de la contribution du béton tendu et de la variation de la direction des contraintes principales de compression sur la hauteur de l'âme. Les résultats sont comparés aux hypothèses de calcul courantes et aux études antérieures, justifiant l'hypothèse habituelle d'un champ de compression uniaxial uniforme dans l'âme. Les régions de discontinuité avec changements brusques de grandeurs statiques sont analysées en utilisant une zone d'appui d'une poutre profilée de hauteur constante comme exemple illustratif. Des champs de contraintes discontinus en forme d'éventails avec résistance à la compression variable du béton sont examinés, et une méthode permettant de vérifier si les contraintes sont au-dessous de la résistance à la compression dans toute la région de l'éventail est présentée. Les résultats sont comparés aux hypothèses courantes de calcul et aux études antérieures. Une méthode de dimensionnement pour les zones d'appui suggérée antérieurement est complétée et justifiée.

La quatrième partie récapitule et discute les résultats obtenus et examine quelques possibilités pour des recherches ultérieures.

Riassunto

La presente tesi vuole contribuire ad una migliore comprensione della capacità portante e della deformazione del cemento armato sottoposto a forze piane normali e di taglio. Vi vengono sviluppati semplici e consistenti modelli fisici in considerazione dei parametri principali, tali da (i) permettere di valutare la capacità di deformazione del cemento armato soggetto ad uno stato di tensione piano; (ii) poter verificare i limiti di applicazione della teoria della plasticità sul cemento armato, e (iii) rivedere, completare ed armonizzare le attuali norme di dimensionamento.

Nella prima parte si esaminano le proprietà essenziali del calcestruzzo e dell'armatura, si riepilogano aspetti sostanziali della teoria della plasticità e della sua applicazione al cemento armato, si passano in rassegna precedenti studi sul cemento armato in stato di tensione piano e si indagano aspetti fondamentali riguardanti il comportamento di lastre di calcestruzzo fessurate.

Nella seconda parte viene presentato un nuovo modello per lastre di calcestruzzo fessurate e armate ortogonalmente, soggette ad uno stato di tensione piano e omogeneo. Nel modello di lastra fessurata si associano i concetti fondamentali dell'approccio originario mediante campi di compressione con un modello di corrente teso recentemente sviluppato. Le distanze tra le fessure e le tensioni tra le stesse vengono determinate partendo da principi meccanici di base; la relazione con i metodi dell'analisi limite è garantita dalla formulazione delle equazioni di equilibrio in termini di tensioni alle fessure, anziché di tensioni medie tra le stesse. Un metodo numerico generale ed una soluzione analitica approssimata vengono derivati e confrontati con precedenti studi teorici e sperimentali. Si propongono semplici espressioni per il carico limite di lastre in cemento armato in funzione del tasso d'armatura e della resistenza a compressione su cilindro del calcestruzzo, si esaminano l'influenza delle forze normali e di precompressione, e vengono infine discussi aspetti essenziali del comportamento di lastre armate in una sola direzione.

Nella terza parte è studiato con modelli semplificati il comportamento al taglio di travi profilate. Viene presentato un procedimento che consente di svolgere analisi di carico-deformazione dell'anima per regioni ove tutte le grandezze statiche e geometriche variano gradualmente lungo l'asse della trave, tenendo conto dell'aderenza delle staffe e della variazione di direzione delle tensioni principali lungo l'altezza della sezione. I risultati vengono comparati con le correnti ipotesi di dimensionamento e con precedenti studi, giustificando la consueta ipotesi di un campo di compressione uniassiale e uniforme nell'anima. Regioni discontinue, caratterizzate da improvvisi cambiamenti delle grandezze statiche, sono analizzate sulla scorta dell'esempio illustrativo della regione di appoggio di una trave profilata di altezza costante. Vengono esaminati campi di tensione discontinui a forma di ventaglio e con una variabile resistenza del calcestruzzo, e si presenta un metodo con cui è possibile la verifica delle tensioni nel calcestruzzo in rapporto alla sua resistenza su tutto il ventaglio. I risultati sono confrontati con le correnti ipotesi di dimensionamento e con studi sperimentali precedenti. Un metodo di dimensionamento per le regioni di appoggio viene completato e giustificato.

Nella quarta parte sono riassunti e discussi i risultati ottenuti e indicati vari possibili sviluppi a livello di ricerche successive.

Table of Contents

1   Introduction

1.1   Defining the Problem

Limit analysis methods have implicitly or explicitly been applied to the solution of engineering problems for a long time. In particular, truss models have been used for following the flow of internal forces in reinforced concrete structures since the very beginning of this construction method. Unfortunately, these methods were thrust into the background for many decades by the emerging theory of elasticity and by empirical and semi-empirical design approaches. However, limit analysis methods were put on a sound physical basis around 1950 through the development of the theory of plasticity and have recently regained the attention of engineers.

Methods of limit analysis provide a uniform basis for the ultimate limit state design of concrete structures. Even for complex problems a realistic estimate of the ultimate load can be obtained with relatively little computational effort. Often, closed form solutions for the ultimate load can be derived; the resulting expressions directly reflect the influences of the governing parameters and the geometry of the problem and give engineers clear ideas of the load carrying behaviour. Theses features are particularly important in conceptual design, where – contrary to refined analyses – all the parameters have to be determined, rather than being known beforehand. Moreover, the theory of plasticity also provides powerful and efficient tools for the dimensioning and detailing of concrete structures. Discontinuous stress fields according to the lower-bound theorem of limit analysis indicate the necessary amount, the correct position and the required detailing of the reinforcement and result in safe designs since the flow of forces is followed consistently throughout the structure.

Application of the theory of plasticity requires sufficient deformation capacity of all structural members and elements. However, while reinforcing steel typically exhibits a rather ductile behaviour, the response of concrete is far from being perfectly plastic. In addition, bond shear stresses transferred between the reinforcement and the surrounding concrete result in a localisation of the steel strains near the cracks, particularly in the post-yield range, reducing the overall ductility of the bonded reinforcement. While sufficient ductility of the reinforcement can usually be ensured by observing appropriate ductility requirements for the reinforcing steel, a ductile behaviour of the concrete can only occasionally be achieved. Hence, it has been argued that limit analysis methods cannot be applied to structural concrete at all. Indeed, the theory of plasticity does not address the questions of the required and provided deformation capacities and thus, additional investigations are required in order to fully justify its application to structural concrete.

1

In design practice one attempts to ensure a sufficient deformation capacity through appropriate detailing measures and usually, the theory of plasticity is applied without deformation checks. Failure governed by concrete crushing is prevented by determining the dimensions from conservative values of the concrete compressive strength. In most cases this approach is adequate from a practical point of view. Collapse of the resulting "underreinforced" structures is governed by yielding of the reinforcement and thus, provided that sufficiently ductile reinforcement is used, the ultimate load according to limit analysis can be achieved. However, in the design of weight-sensitive structures such as long-span bridges or offshore platforms as well as in the increasingly important area of the evaluation of existing structures, the concrete dimensions cannot be liberally increased. Furthermore, the application of modern high-strength concrete cannot be justified if most of its beneficial strength is lost due to excessively conservative assumptions. Finally, the approach outlined above is certainly not satisfactory from a more fundamental point of view. Uncertainties frequently arise when attempting to establish whether and how deformations should be checked and often, the application of the theory of plasticity is limited by excessive restrictions, counteracting the engineering ideals of structural efficiency and economy.

The reason for these difficulties lies in the fact that at present, no consistent and experimentally verified theory of the deformation capacity of structural concrete is available. This thesis is part of the research project 'Deformation Capacity of Structural Concrete' which aims at developing such a theory that will allow one (i) to discuss questions of the demand for and the supply of deformation capacity, (ii) to explore the limits of applicability of the theory of plasticity to structural concrete, and (iii) to critically review, supplement and harmonise current design provisions. Previous work within the overall project includes several series of large-scale tests, an examination of the deformation capacity of structural concrete girders [144] as well as an investigation focusing on the influence of bond behaviour on the deformation capacity of structural concrete [6].

This thesis covers the behaviour of structural concrete subjected to in-plane shear and normal forces. Apart from a wide range of limit analysis methods, previous work on plane stress in structural concrete includes compression field approaches that allow predicting complete load-deformation curves. Basically, such approaches would be suitable for a discussion of the questions of the required and provided deformation capacities. However, in previous approaches tension stiffening effects were either neglected, resulting in much too soft response predictions, or they were accounted for by empirical constitutive equations relating average stresses and average strains in tension. While a better match with experimental data could be obtained from such modified approaches, the direct link to limit analysis was lost. Moreover, the underlying empirical constitutive equations relating average stresses and average strains in tension are debatable and do not yield information on the maximum steel and concrete stresses at the cracks nor on the amount of strain localisation in the reinforcement near the cracks. Hence, based on the existing approaches no satisfactory assessment of the deformation capacity of structural concrete subjected to in-plane loading is possible.

1.2 Scope

This thesis aims at contributing to a better understanding of the load-carrying and deformational behaviour of structural concrete subjected to in-plane shear and normal forces, including membrane elements (homogeneous state of plane stress) and webs of girders with profiled cross-section (non-homogeneous state of plane stress). Simple, consistent physical models reflecting the influences of the governing parameters shall be developed on whose basis (i) a realistic assessment of the deformation capacity of structural concrete subjected to in-plane loading is possible, (ii) the limits of applicability of the theory of plasticity to structural concrete can be explored, and (iii) current design provisions can be critically reviewed, supplemented and harmonised.

Furthermore, existing models for the behaviour of structural concrete subjected to in-plane shear and normal stresses shall be reviewed in order to clarify the underlying assumptions and the relationships between the different approaches.

1.3 Overview

In the first part of this thesis material properties are examined, fundamental aspects of the theory of plasticity are summarised, and previous work on plane stress problems is reviewed. Chapter 2 examines the behaviour of concrete, reinforcement and their interaction, focusing on simple physical models reflecting the main influences governing the response of structural concrete. Chapter 3 summarises the theory of plastic potential for perfectly plastic materials and discusses the basic aspects of its application to structural concrete. Chapter 4 investigates the fundamental aspects of the behaviour of cracked concrete membranes, reviews previous work on plane stress problems, describes the relationships between the different approaches and clarifies the underlying assumptions.

The second part, Chapter 5, covers the behaviour of membrane elements and presents a new model for cracked, orthogonally reinforced concrete panels subjected to a homogeneous state of plane stress. Both a general numerical method and an approximate analytical solution are derived and the results are compared with previous theoretical and experimental work, including a detailed comparison with limit analysis methods. The influences of prestressing and axial forces are examined and basic aspects of the load-carrying behaviour of uniaxially reinforced membrane elements are discussed.

In the third part, Chapter 6, the behaviour of beams in shear is examined, focusing on simplified models for girders with flanged cross-section. Chapter 6.2 investigates situations where all static and geometric quantities vary only gradually along the girder axis. An approximate model for the load-deformation behaviour of the web is derived and the results are compared with typical design assumptions and with previous experimental work. In Chapter 6.3 discontinuity regions characterised by abrupt changes of static quantities are analysed using fan-shaped discontinuous stress fields. A method that allows checking whether the concrete stresses are below the concrete compressive strength throughout the fan region is presented, accounting for the degradation of the concrete compressive strength due to lateral tensile strains. The results are compared with typical design assumptions and with previous experimental work.

The fourth part, Chapter 7, summarises and discusses the results obtained in the first three parts of this thesis and concludes with a set of recommendations for future research.

1.4 Limitations

Throughout this thesis only small deformations are considered, such that changes of geometry at the ultimate state are insignificant and hence, the principle of virtual work can be applied to the undeformed members. Short-term static loading is assumed, excluding dynamic or cyclic loads as well as long-term effects.

Apart from a brief examination of the load-carrying behaviour of uniaxially reinforced membrane elements only orthogonally reinforced members are treated throughout Chapters 5 and 6, assuming rotating, stress-free, orthogonally opening cracks. Consideration of the uncracked behaviour is excluded, and the elements or webs, respectively, are assumed to be of constant thickness and provided with a minimum reinforcement capable of carrying the applied stresses at cracking. Aspects of fracture mechanics are only covered on the material level, Chapter 2, and neither fibre nor non-metallic reinforcement is considered.

In girders, only the portion of the shear force carried by the web is considered, excluding contributions of the flanges to the shear resistance. In the numerical examples typical distributions of the chord strains are assumed, neglecting possible interactions between the state of stress in the web and the chord strains. Furthermore, only some basic aspects of the influence of curved prestressing tendons are discussed.

It should be noted that tensile stresses and strains are taken as positive throughout this thesis.

2 Material Properties

2.1 General Considerations

In this chapter, the properties of concrete and reinforcement relevant for structural concrete subjected to in-plane stresses are examined. Existing models for the behaviour of concrete, typically established on the basis of tests on low- and normal-strength specimens, are compared with recent tests on high-strength concrete specimens. Such a comparison is appropriate since concretes in common use today have considerably higher strengths than concretes produced some years ago. Though not of primary interest for structural concrete subjected to in-plane stresses, test results of triaxially compressed and confined concrete of different strengths are also included.

Rather than attempting to provide a complete mechanical description of the behaviour of concrete, reinforcement and their interaction, physical models are aimed at which are as simple as possible and reflect the main influences governing the response of structural concrete. Much of the work presented in this chapter is based on a report by Sigrist [144], who gave a detailed description of many of the models adopted, in particular for the confinement effect in columns and the strain-softening behaviour of concrete in tension and compression.

The diagrams shown in Fig. 2.1 illustrate some basic aspects and possible idealisations of stress-strain characteristics. The response shown in Fig. 2.1 (a) is (non-linear) elastic; there is a unique relationship between strains and applied stresses, the deformations are completely reversible, and no energy is dissipated. The strain energy per unit volume, corresponding to the energy stored in an elastic body, is given by

$$dU = \int_{\varepsilon} \sigma(\varepsilon) d\varepsilon \tag{2.1}$$

and represented by the shaded area below the stress-strain curve in Fig. 2.1 (a). The shaded area above the stress-strain curve corresponds to the complementary strain energy per unit volume, defined as

$$d\overset{*}{U} = \int_{\sigma} \varepsilon(\sigma) d\sigma \tag{2.2}$$

At any point of the stress-strain curve, the sum of the strain energy dU and the complementary strain energy $d\overset{*}{U}$ per unit volume equals $\sigma \cdot \varepsilon$; for linear elastic behaviour, both energies are equal to $\sigma \cdot \varepsilon / 2$.

Fig. 2.1 (b) shows an elastic-plastic stress-strain relationship; the deformations are not fully reversible. Upon unloading, only the portion of the strain energy below the unloading curve is released. The remaining energy dD, corresponding to the area between the loading and unloading curves, has been dissipated. Strain-hardening branches of stress-strain curves are characterised by irreversible deformations and energy dissipation under increasing loads and deformations. Strain-softening branches of stress-strain curves, exhibiting decreasing loads with increasing deformation, Fig. 2.1 (c), can only be recorded by means of strict deformation control. Generally, the strain-softening branch of a stress-strain diagram not only reflects the material behaviour but the response of the entire structural system including effects from the testing machine; further explanations are given in Chapters 2.2.1 and 2.2.2.

Figs. 2.1 (d)-(f) illustrate some commonly used idealisations of stress-strain relationships. In the bilinear representation shown in Fig. 2.1 (d), the response is linear elastic, $\sigma = E\varepsilon$, for stresses below the yield stress, where E = modulus of elasticity. For higher stresses, $\sigma > f_y$, a linear strain-hardening takes place, $d\sigma = E_h d\varepsilon$, where E_h = hardening modulus; unloading is assumed to occur parallel to the initial elastic loading. By adapting the parameters E, E_h and f_y a bilinear model can be used to closely approximate most stress-strain characteristics observed in tests, apart from the post-peak range. If only ultimate loads and initial stiffnesses are of interest, a linear elastic-perfectly plastic

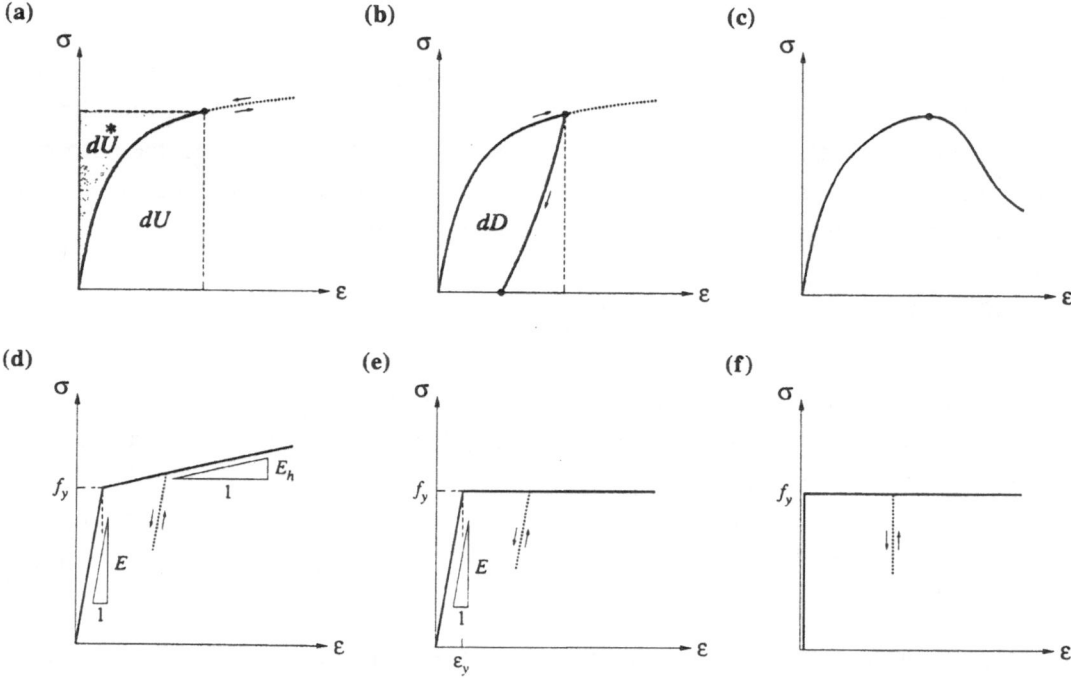

Fig. 2.1 – Stress-strain characteristics: (a) elastic response; (b) elastic-plastic response; (c) strain-softening behaviour; (d) bilinear, (e) linear elastic-perfectly plastic, and (f) rigid-perfectly plastic idealisations.

idealisation of the stress-strain response may be adequate, Fig. 2.1 (e). A simple rigid-perfectly plastic idealisation, Fig. 2.1 (f), is often sufficient in the assessment of ultimate loads. Rigid- and linear elastic-perfectly plastic behaviour can be regarded as special cases of the bilinear idealisation.

As shown in Chapter 5 an adequate description of the basic mechanisms of interaction between concrete and reinforcement provides the key to a better understanding of the behaviour of structural concrete subjected to in-plane stresses. Therefore, an appropriate model for bond and tension stiffening is essential for the present work. Another important aspect is the behaviour of structural concrete subjected to biaxial compression and tension which, contrary to the biaxial behaviour of plain concrete, is still a matter of disagreement among researchers, in spite of numerous theoretical and experimental investigations on the dependence of the concrete compressive strength on transverse tensile stresses and strains. An attempt will be made here to overcome the apparent discrepancies in the available test data.

Material properties determined from tests depend on the particular testing method used. Therefore, to allow for a direct comparison of test results standardised testing methods (including specimen geometry, loading ratio and testing device) should be applied. Unfortunately, this is frequently not the case, and some of the scatter observed when comparing test results obtained by different researchers has to be attributed to this situation.

2.2 Behaviour of Concrete

2.2.1 Uniaxial Tension

The tensile strength of concrete is relatively low, subject to rather wide scatter and may be affected by additional factors such as restrained shrinkage stresses. Therefore, it is common practice to neglect the concrete tensile strength in strength calculations of structural concrete members. However, this is not always possible; e.g., the shear resistance of girders without stirrups depends on tensile stresses in the concrete. Furthermore, the tensile behaviour of concrete is a key factor in serviceability considerations such as the assessment of crack spacings and crack widths, concrete and reinforcement stresses and deformations.

Basically, the tensile strength of concrete can be determined from direct tension tests, Fig. 2.2 (a). However, such tests are only rarely used, even in research, because of the difficulties to achieve truly axial tension without secondary stresses induced by the holding devices. Usually, the concrete tensile strength is evaluated by means of indirect tests such as the bending or modulus of rupture test, Fig. 2.2 (b), the double punch test, Fig. 2.2 (c), or the split cylinder test, Fig. 2.2 (d). While easier to perform, indirect tests require assumptions about the state of stress within the specimen in order to calculate the tensile strength from the measured failure load. For most purposes, an estimate of the tensile strength based on the uniaxial compressive strength is sufficient; $f_{ct} = 0.3(f_c')^{2/3}$ in MPa may be assumed as an average value for normal strength concrete, where f_c' is the cylinder compressive strength of concrete.

The stress-strain response of a concrete member in uniaxial tension, Figs. 2.3 (a) and (b), is initially almost linear elastic. Near the peak load the response becomes softer due to microcracking, and, as the tensile strength is reached, a crack forms. However, the tensile stress does not instantly drop to zero as it would in a brittle material like glass. Rather, the carrying capacity decreases with increasing deformation, i.e. a strain-softening or quasi-brittle behaviour can be observed. The capability of concrete to transmit tensile stresses after cracking may be attributed to bridging by aggregate particles [64]. This aspect of the behaviour of concrete has been known for only about 25 years because very stiff testing machines and highly sensitive and precise measuring devices are necessary in order to record the post-peak behaviour of concrete in tension.

Tests show that the softening branch of the stress-strain diagram of longer specimens is steeper than that of shorter specimens, Fig. 2.3 (b), and for specimens longer than a certain critical length, the softening branch cannot be recorded at all. The fact that long specimens fail in a more brittle manner than short ones cannot be explained by continuum mechanics models like a stress-strain diagram. Due to the quasi-brittle nature of concrete, linear elastic fracture mechanics cannot be applied either, except for infinitely large specimens [64].

Hillerborg [53] introduced the "fictitious crack model" which is capable of describing the failure of concrete in tension. After the peak load has been reached, the parts of the member away from the crack unload, Fig. 2.3 (c), and the deformations of the member localise at the crack or in its vicinity, the so-called fracture process zone. This development is called strain localisation. Considering a fictitious crack, i.e., a fracture process zone of zero initial length, fracture behaviour can be described by a stress-crack opening relationship, Fig. 2.3 (d). The area below the stress-crack opening curve represents the specific fracture energy in tension G_F, dissipated per unit area of the fracture process zone until complete separation of the specimen has occurred. If G_F is assumed to be a

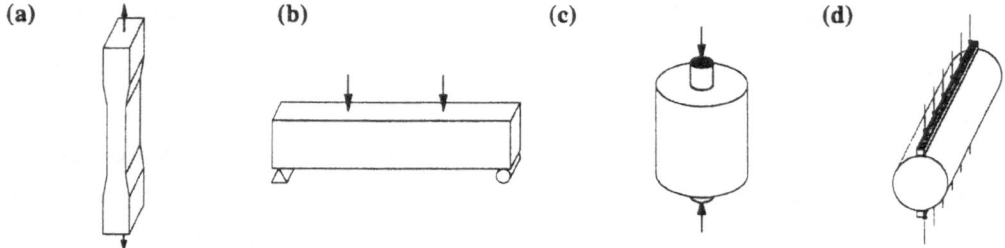

Fig. 2.2 – Tension tests: (a) direct tension test; (b) bending or modulus of rupture test; (c) double punch test; (d) split cylinder test.

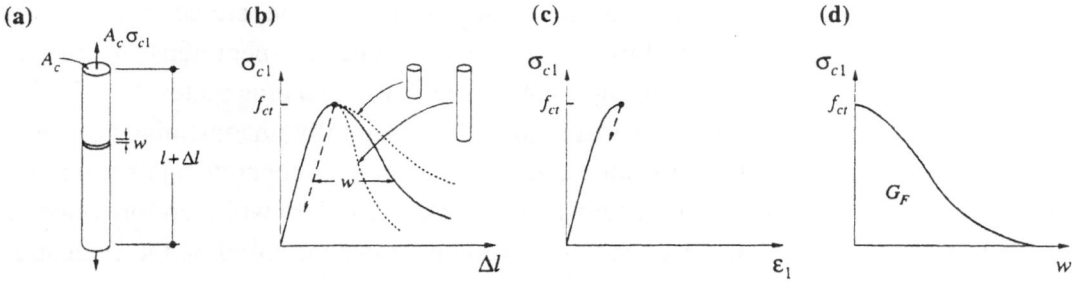

Fig. 2.3 – Fictitious crack model: (a) test specimen; (b) influence of specimen length; (c) stress-strain diagram for regions outside the fracture process zone; (d) stress-crack opening relationship of fictitious crack.

material property, the dependence of the post-peak behaviour on specimen length can be explained since the stored energy increases in proportion to specimen length, while the energy dissipated at failure remains constant at $A_c G_F$. Also, high-strength concrete specimens fail in a more brittle manner since G_F only slightly increases with the concrete strength; more details are given in [144]. Instead of assuming a fictitious crack of zero initial length and a stress-crack opening relationship, a fracture process zone of finite length can be assumed along with a stress-strain relationship for this process zone. This approach is called "crack band model" and is equivalent to the fictitious crack model if both the initial length as well as the stress-strain relationship of the fracture process zone are assumed to be material properties. The crack band model is more suitable for finite element applications; for more details and a comprehensive survey of other fracture mechanics approaches see [64].

2.2.2 Uniaxial Compression

The response of concrete in uniaxial compression is usually obtained from cylinders with a height to diameter ratio of 2, Fig. 2.4 (a). The standard cylinder is 300 mm high by 150 mm in diameter, and the resulting compressive cylinder strength is termed f_c'. Smaller size cylinders and cubes, Fig. 2.4 (b), are often used for production control, the latter mainly because such tests do not require capping or grinding of the specimen ends. When evaluating test results it is important to note that strengths measured on smaller cylinders and cubes are typically higher than those determined from standard cylinders since the end zones of the specimens are laterally constrained by the stiffer loading plates, an effect more pronounced in small specimens and particularly in cubes. The difference between the cube strength f_{cc} and the cylinder strength f_c' decreases with increasing concrete strength; approximate relationships are given in Fig. 2.4 (c).

Uniaxial compression tests on wall elements of plain concrete result in strengths about 10...20% lower than tests on standard cylinders; this can be attributed to the different failure modes observed in these tests, Figs. 2.5 (b) and (c). While laminar splitting failures, i.e., cracks forming parallel to the compressive direction, are common in wall elements, Fig. 2.5 (c), sliding failure is observed in cylinder specimens of normal-strength concrete since laminar splitting is constrained by the loading plates, Fig. 2.5 (b). The observation that the compressive strength f_c of a laterally unconstrained concrete element is lower than f_c' thus indicates that the resistance of concrete against laminar splitting is lower than its resistance against sliding; sliding failure will therefore only occur if additional resistance against laminar splitting is provided. Based on the evaluation of many test results, Muttoni et al. [107] proposed

$$f_c = 2.7(f_c')^{2/3} \text{ in MPa,} \qquad \text{where} \qquad f_c \leq f_c' \qquad (2.3)$$

for the compressive strength f_c of a laterally unconstrained concrete in uniaxial compression. According to Eq. (2.3), f_c increases less than proportional with f_c'; a possible explanation for this behaviour follows again from the observation of failure modes, Fig. 2.5 (f). High-strength concrete cylinders often fail by laminar splitting although the specimen ends are constrained, i.e., additional resistance against laminar splitting provided by the specimen ends is not enough to induce sliding failure in high-strength concrete specimens. This indicates that the difference between the resistances against sliding and laminar splitting, and therefore the difference between f_c' and f_c, increases with concrete strength. The compressive strength of concrete in a structural concrete element depends on additional parameters; more details are given in Chapter 2.4.3.

The uniaxial compressive strength is often the only concrete property specified and measured. The compressive stress-strain response of concrete in the pre-peak range can be approximated by a parabola, Fig. 2.5. (a),

$$\frac{\sigma_{c3}}{f_c} = \frac{(\varepsilon_3^2 + 2\varepsilon_3\varepsilon_{co})}{\varepsilon_{co}^2} \qquad (2.4)$$

where ε_{co} = concrete strain at peak compressive stress f_c. The value of ε_{co} is almost constant at $\varepsilon_{co} \approx 0.002$ for normal-strength concrete ($f_c \leq 30$ MPa); for higher concrete strengths, a slight increase to about $\varepsilon_{co} \approx 0.003$ at $f_c = 100$ MPa has been observed. While Eq. (2.4) closely approximates the response of normal-strength concrete, the

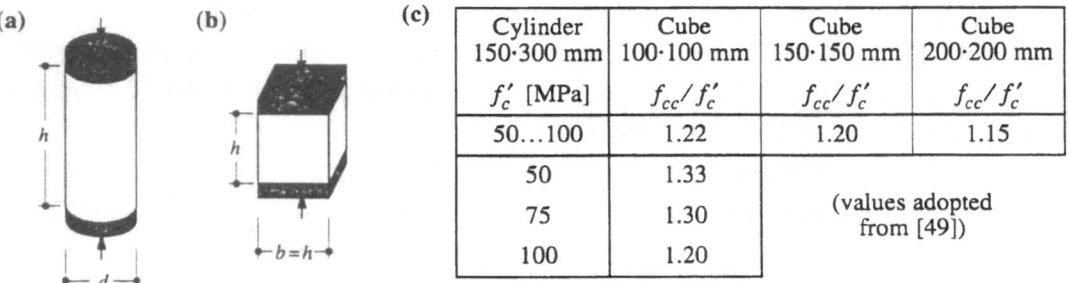

(a)	(b)	(c) Cylinder 150·300 mm	Cube 100·100 mm	Cube 150·150 mm	Cube 200·200 mm
		f_c' [MPa]	f_{cc}/f_c'	f_{cc}/f_c'	f_{cc}/f_c'
		50...100	1.22	1.20	1.15
		50	1.33		
		75	1.30	(values adopted from [49])	
		100	1.20		

Fig. 2.4 – Compression tests: (a) cylinder test; (b) cube test; (c) effect of specimen size and geometry on measured compressive strength.

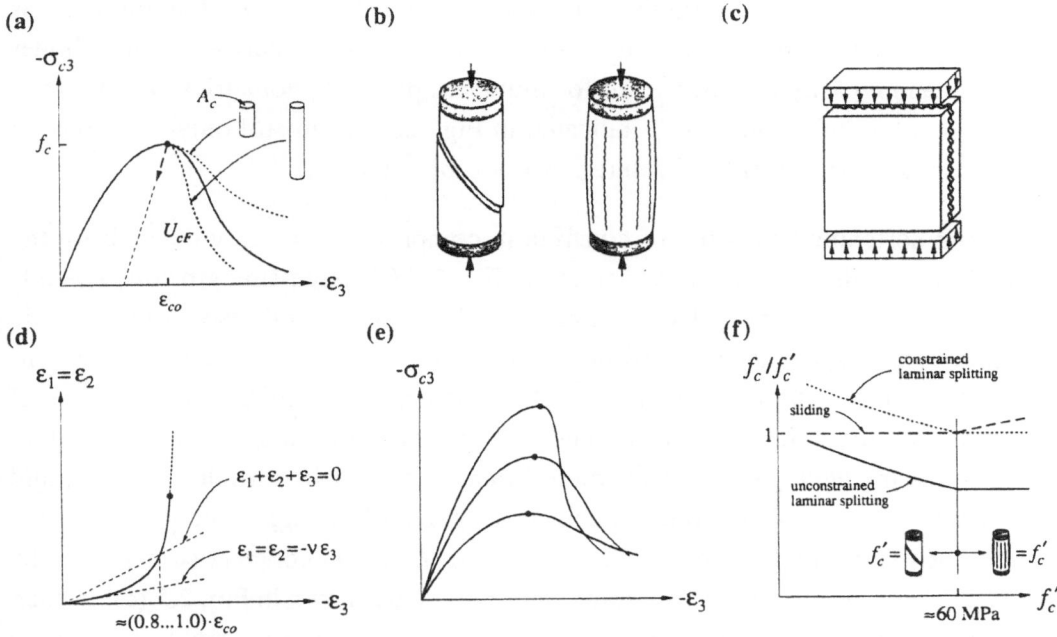

Fig. 2.5 – Uniaxial compression: (a) stress-strain response and influence of specimen length on strain-softening; (b) and (c) failure modes; (d) axial and lateral strains; (e) influence of concrete strength on strain-softening; (f) assumed influence of cylinder strength on resistance against different failure modes.

stress-strain relationships of high-strength concrete are initially almost linear and less curved than predicted by Eq. (2.4).

Fig. 2.5 (d) shows the development of lateral strains in an axisymmetrical specimen loaded in uniaxial compression according to Fig. 2.5 (a). Initially, the response is approximately linear elastic, $\varepsilon_1 = \varepsilon_2 = -\nu\varepsilon_3$. Already at a comparatively low compressive stress of $\sigma_{c3} \approx -f_c/3$, lateral strains start to increase more rapidly due to microcracking until shortly before failure, at axial stresses of about $\sigma_{c3} = -(0.8...1.0)\,f_c$, the volumetric strain $\varepsilon_v = \varepsilon_1 + \varepsilon_2 + \varepsilon_3$ becomes positive, i.e., the specimen dilates. A similar behaviour is observed in multiaxial tests. The volumetric strain, which measures the volumetric expansion and thereby the degree of damage of a material, has therefore been considered as the state variable governing the failure of concrete [119].

Similar to the behaviour in uniaxial tension, the response of concrete in compression in the post-peak range is characterized by decreasing carrying capacity with increasing deformation, i.e. strain-softening in compression, Fig. 2.5 (a). As in uniaxial tension, the softening branch of long specimens is steeper than that of short specimens, which may again be attributed to the localisation of deformations in a fracture process zone, while the remaining parts of the specimen are unloaded. However, the strain-softening behaviour of concrete in compression is more complicated than that in tension, and no generally accepted model such as the fictitious crack model for the behaviour in tension has yet been established. One reason for this is that the size and shape of the fracture process zone, which may be assumed to extend over a length of approximately $l = 2d$ in cylinder tests [144], cannot easily be determined for more complicated geometries. The specific fracture energy per unit volume U_{cF} indicated in Fig. 2.5 (a) can still only be evaluated from test results if the size of the fracture process zone is known.

The strain-softening behaviour of concrete in compression not only depends on the specimen size, but also on the concrete strength, Fig. 2.5 (e). High-strength concrete fails in a much more brittle manner than normal-strength concrete; while the specific elastic energy stored in the specimen is proportional to f_c^2, the specific fracture energy U_{cF} increases only slightly with the concrete strength. This may be attributed to the change of failure modes observed for concrete strengths of $f_c' \approx 50$ MPa; above this value failure occurs through the aggregate particles rather than at the matrix-particle interface, and thus the fracture energy of concrete is controlled by that of the aggregate particles. Another possible explanation for the modest increase of U_{cF} with concrete strength can be derived from considering the laminar splitting failure mode shown in Fig. 2.5 (c). In such failures, the specific fracture energy in compression U_{cF} can be expected to be proportional to the length of the fracture process zone and to the specific fracture energy in tension of the laminar cracks, G_F, which increases only slightly with f_c', Chapter 2.2.1. The proportionality of U_{cF} to G_F can also be assumed if failure occurs by a combination of laminar splitting and sliding [79].

2.2.3 Biaxial Loading

The behaviour of plain concrete in plane stress has been investigated by many researchers. Kupfer [73] published the first reliable test results, using concrete strengths of $f_c' = 19...60$ MPa. He tested 228 specimens loaded by means of steel brushes of calibrated stiffness – eliminating confinement of the specimen ends as far as possible – as well as 24 specimens loaded through solid steel plates. The latter tests confirmed that most of the strong increase of concrete strength in biaxial compression observed in earlier tests was due to confinement of specimen ends by the loading plates. The results of the tests without confinement of the specimen ends are shown in Fig. 2.6 along with more recent test results obtained by other researchers, all from specimens loaded by means of steel brushes. Biaxial strengths are given with respect to the uniaxial compressive strength of specimens identical to those used in the biaxial tests. Interaction relationships proposed by Kupfer [73] and Nimura [114] are also indicated, along with the square failure criterion suggested by several researchers; the latter completely neglects the tensile strength as well as any increase of strength in biaxial compression. More details on failure criteria are given in Chapter 3.2.3.

The test results of Nimura [114] and his proposed interaction relationship indicate a less pronounced increase of strength in biaxial compression at $\sigma_1 \approx \sigma_2$ for high-strength concrete, while small lateral compressive stresses appear to have a more beneficial ef-

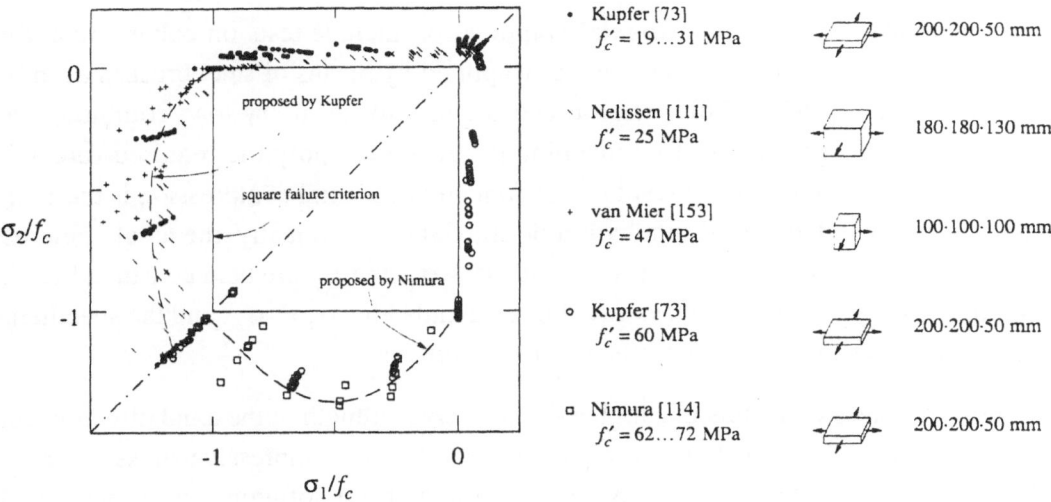

Fig. 2.6 – Biaxial concrete strength: test results of normal-strength (upper left) and high-strength concrete (lower right).

fect; a similar behaviour has been observed on smaller specimens by Chen et al. [26]. On the other hand, Kupfer [73] investigated a wide range of concrete strengths and did not find significant differences in behaviour. The observed discrepancies might simply be due to differences in the testing machines used by the different researchers.

Depending on the ratio σ_1/σ_2 of the applied stresses, where $\sigma_1 \geq \sigma_2$ (compression negative), different failure modes can be distinguished; note that the boundaries of the failure regimes given are subject to rather wide scatter. In biaxially compressed specimens with $\sigma_1/\sigma_2 = 1...0.3$, laminar splitting failures are observed, Fig. 2.5 (c). Laminar splitting combined with cracks parallel to the direction of σ_2 occurs for biaxial compression with $\sigma_1/\sigma_2 < 0.3$ and for tension-compression with small tensile stresses of about $\sigma_1/\sigma_2 \geq -0.05$. Finally, for tension-compression with higher tensile stresses, and for biaxial tension, tensile failures occur, i.e., cracks form parallel to the direction of σ_2. The sliding mode of failure observed in cylinder tests is rarely encountered if confinement of the specimen ends is prevented.

After a tensile failure in biaxial tension-compression has occurred, compressive stresses parallel to the cracks can still be transmitted. If the compressive stress is further increased, the specimen will eventually fail in compression at a load approximately equal to the uniaxial compressive strength. For small tensile stresses, failure in biaxial tension-compression occurs by laminar splitting, generally at a compressive stress somewhat lower than the uniaxial compressive strength. The compressive strength of concrete may thus be reduced by lateral tensile stresses, even in structural concrete where tensile stresses can be transferred to the reinforcement upon cracking (Chapter 2.4.3).

2.2.4 Triaxial Compression

Testing methods for concrete in triaxial compression include tests on cubes and cylinders. In tests on cubes, the loads are typically applied by means of steel brushes as in biaxial compression [153]. While tests on cubes allow for arbitrary load combinations, $\sigma_1 \neq \sigma_2 \neq \sigma_3$, a complicated testing machine is needed to apply the required forces in three directions. More often, the response of concrete in triaxial compression is therefore obtained from cylinders tested in a hydraulic triaxial cell. Typically, the axial compressive stress $-\sigma_3$ is increased while the radial stresses σ_1 and σ_2 are held constant; the lateral compressive stresses are always equal in a triaxial cell, $-\sigma_1 = -\sigma_2$, and the specimens are coated in order to avoid pore pressures in the concrete.

Lateral compression results in a higher compressive strength in the axial direction and a greatly enhanced ductility. Even at relatively small lateral compressive stresses, strains at the ultimate state are substantially increased and strain-softening in the post-peak range is much less pronounced; the only aspect that will be further examined here is the increase of the triaxial compressive strength due to lateral compression. Test results indicate that for moderate lateral compressive stresses of about $-\sigma_1 \leq 2 f_c$, the triaxial compressive strength f_{c3} increases by roughly four times the applied lateral compressive stress, i.e.,

$$f_{c3} = f_c - 4\sigma_1 \tag{2.5}$$

A relation of this type has already been proposed by Richart et al. [127], using a proportionality factor of 4.1 in the last term of Eq. (2.5). Fig. 2.7 shows test results obtained from cylinders 200 mm high by 100 mm in diameter compared to Eq. (2.5); the agreement is satisfactory, with no significant differences between normal and high-strength concrete. Note that for high-strength concrete, only a limited range of lateral pressures has been investigated since the ultimate loads at higher lateral pressures would exceed the capacity of the testing machine.

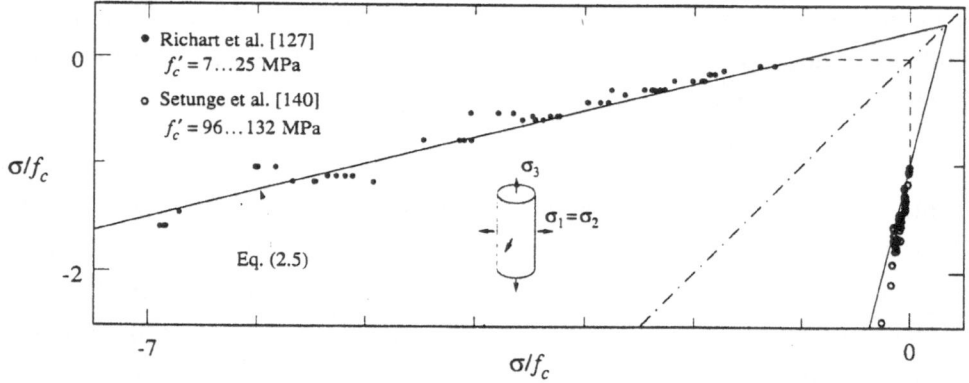

Fig. 2.7 – Triaxial compression: test results of normal-strength (upper left) and high-strength concrete (lower right).

If – as in uniaxial compression – the volumetric strain is considered as the state variable governing the failure of concrete, the increase of the triaxial strength might have to be attributed to the lateral restraint stiffness rather than the lateral compressive stress [119]. Since most triaxial tests have been conducted in similar devices, this question cannot be settled based on the available test data.

2.2.5 Aggregate Interlock

The ability of concrete to transmit stresses across cracks is termed aggregate interlock. Aggregate interlock is particularly important in connections of precast concrete segments and in plane stress situations if the principal stress directions change during the loading process. Much theoretical and experimental work has been done in order to establish aggregate interlock relationships between the crack displacements, δ_n and δ_t, and the normal and shear stresses σ_n and τ_{tn} acting on the crack surface, Fig. 2.8 (a). While earlier research focused on the bearing capacity of connections in precast concrete construction, leading to the shear friction analogy established among others by Birkeland [15], Mast [97] and Mattock [55,98,99], later investigations tried to develop complete constitutive relationships of crack behaviour [11,162,132,36]. A comprehensive review of research in the field of aggregate interlock can be found in [115].

Tests of aggregate interlock behaviour are demanding since very small displacements of irregular crack faces have to be controlled under large forces. Specimens often fail away from the cracks, and due to the irregular crack faces, the state of stress in the crack area is hardly ever uniform, even in more complicated test setups [38] than the commonly used push-off test illustrated in Fig. 2.8 (b). Test results are therefore subject to wide scatter. Lateral restraint of push-off specimens can be provided by passive external steel bars or by a system with actuators, allowing to keep lateral forces or lateral deformations constant during the test. Alternatively, internal transverse reinforcement can be used.

Analytically, an aggregate interlock relationship can be expressed by a crack stiffness matrix $K^{(r)}$ relating crack displacements, δ_n and δ_t, to the stresses σ_n and τ_{tn}. The quantities δ_n, δ_t, σ_n, and τ_{tn} should be considered as average values over several cracks and large crack areas because of the irregular nature of the crack surfaces. An appropriate description of aggregate interlock behaviour would require an incremental formulation,

$$\begin{bmatrix} d\sigma_n \\ d\tau_{tn} \end{bmatrix} = K^{(r)} \begin{bmatrix} d\delta_n \\ d\delta_t \end{bmatrix} \qquad \text{where} \qquad K^{(r)} = K^{(r)}(\delta_n, \delta_t, \sigma_n, \tau_{tn}, ...) \qquad (2.6)$$

since the crack stiffness matrix $K^{(r)}$ generally depends on δ_n, δ_t, σ_n, τ_{tn} and the loading history, i.e., the behaviour is path-dependent. Some qualitative requirements for an adequate aggregate interlock relationship, Eq. (2.6), follow from theoretical considerations [11,83]: the crack opening cannot be negative, $\delta_n \geq 0$; the normal stresses on the crack cannot be tensile, $\sigma_n \leq 0$; for a pure crack opening, $d\delta_n > 0$, $d\delta_t = 0$, shear and normal stresses must decrease, $d\sigma_n \leq 0$, $d\tau_{tn} \leq 0$ (or, in a rigid-plastic material, $\sigma_n = 0$, $\tau_{tn} = 0$

because the crack faces are not in contact). The crack stiffness matrix is generally not positive definite, indicating that crack behaviour on its own is unstable and cannot be modelled by linear or non-linear springs. Based on such theoretical considerations and noting that available experimental data do not allow establishing a relation of the general type, Eq. (2.6), Bazant and Gambarova [11] neglected path-dependence and proposed a simpler relation of the type $\sigma_n = \sigma_n(\delta_n, \delta_t)$, $\tau_{tn} = \tau_{tn}(\delta_n, \delta_t)$. Their "rough crack model" has been adopted by several researchers (Chapter 4.3.3).

Neglecting path-dependence as well, Walraven [161,162] established a physically-based aggregate interlock relationship, Fig. 2.8 (c), in which aggregate interlock stresses are evaluated from randomly distributed spherical, rigid aggregate particles of various size penetrating a rigid-perfectly plastic mortar matrix as

$$\sigma_n = \sigma_n(\delta_n, \delta_t) = f_{my}(A_t - \mu A_n)$$
$$\tau_{tn} = \tau_{tn}(\delta_n, \delta_t) = f_{my}(A_n + \mu A_t)$$

(2.7)

where f_{my} and μ are the yield strength and the coefficient of friction of the rigid-perfectly plastic mortar matrix, and A_t and A_n are statistically evaluated integrals of the projections of the contact surfaces, a_t and a_n, Fig. 2.8 (c). A_t and A_n generally depend on the crack displacements, δ_n and δ_t, as well as on the maximum aggregate diameter and the volume fraction of aggregate per unit volume of concrete; the values of f_{my} and μ were determined from a comparison with numerous tests. Walraven observed that the behaviour of cracks in reinforced concrete (specimens laterally restrained by internal reinforcement) differed significantly from the behaviour of cracks in plain concrete; the difference was such that it could not be explained by dowel action of the reinforcement. Cracks in reinforced concrete typically showed a much stiffer response, and the crack opening direction happened to be almost the same in all specimens of this type [161]. Walraven concluded that due to bond stresses, the crack width in the vicinity of the reinforcement was smaller than the average crack width, and therefore, compressive diagonal struts formed near the reinforcement, resulting in a stiffer response and higher ultimate loads. Nevertheless, Walraven's aggregate interlock relationship, Eq. (2.7), has been adopted by

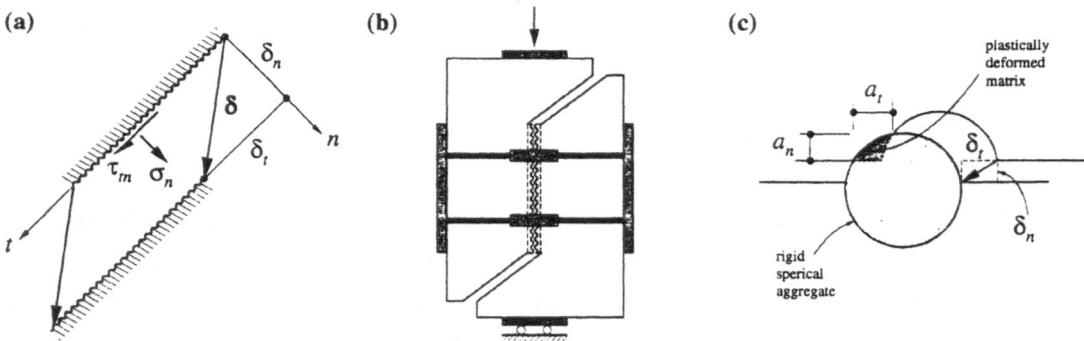

Fig. 2.8 – Aggregate interlock: (a) notation; (b) push-off specimen; (c) matrix-aggregate interaction according to Walraven's model [161,162].

many researchers in order to describe the behaviour of cracks not only in plain, but also in reinforced concrete, since it is the only consistent physical model available.

Based on Walraven's work and a comparison with limit analysis solutions, Brenni [17] proposed a simplified aggregate interlock relationship

$$\tau_{tn} = \frac{\sqrt{\sigma_n(f_c - \sigma_n)}}{1 + \delta_n/c_1}, \qquad \sigma_n = \frac{f_c}{c_2}\left(\frac{\delta_n}{\delta_t} - c_3\right) \tag{2.8}$$

which agrees well with Walraven's experimental results for $c_1 = 0.8$ mm, $c_2 = 15$ and $c_3 = 0.6$. For $\delta_n = 0$, Eq. $(2.8)_1$ yields the shear strength governed by concrete crushing according to limit analysis, Chapter 4.2.

2.3 Behaviour of Reinforcement

2.3.1 General

The use of iron in order to reinforce concrete structures dates back to the end of the last century and marks the birth of reinforced concrete construction. In the beginning, there were several reinforcement systems, using different shapes and types of iron or steel. Common reinforcement types today are deformed steel bars of circular cross-section for passive reinforcement and steel bars, wires or seven-wire strands for prestressed reinforcement. The deformation capacity of structural concrete elements, an important aspect in the design of such structures [144], mainly depends on the ductility of the reinforcement, and structural concrete elements are generally designed such that failure will be governed by yielding of the reinforcement. Therefore, ductility of the reinforcement is as essential to structural concrete as its strength.

Much research has been conducted over the past decades in the field of non-metallic reinforcement, including glass, carbon and aramid fibres. Randomly distributed glass fibres result in smaller crack widths and hence, better serviceability. If glass fibres alone are used as reinforcement, very high quantities of fibres are required in order to achieve desired resistances, and the workability of the concrete-glass fibre mix becomes troublesome. Glass fibres are therefore mainly used for crack-control in prefabricated non-structural elements; steel fibres can also be applied for this purpose, but they are less suitable due to corrosion problems. Carbon and aramid fibres have higher strengths than steel while their weight is considerably lower, and they do not corrode; such materials are potentially interesting for use in long-span structures, preferably as prestressing cables. However, such fibres are brittle, i.e., their response in axial tension is almost perfectly linear elastic until rupture, and they are sensitive to lateral forces, which complicates their application. In addition, carbon and aramid fibres are relatively expensive. A 50 m carbon or aramid fibre post-tensioning cable, including anchors, is 3...10 times more expensive today than a steel cable of equal resistance, and 6...25 times more than one of equal stiffness [164].

2.3.2 Reinforcing Steel

Two basically different types of stress-strain characteristics of reinforcing steel can be distinguished. The response of a hot-rolled, low-carbon or micro-alloyed steel bar in tension, Fig. 2.9 (a), exhibits an initial linear elastic portion, $\sigma_s = E_s \varepsilon_s$, a yield plateau (i.e., a yield point at $\sigma_s = f_{sy}$ beyond which the strain increases with little or no change in stress), and a strain-hardening range until rupture occurs at the tensile strength, $\sigma_s = f_{su}$. Various steel grades are usually defined in terms of the yield strength f_{sy}. The extension of the yield plateau depends on the steel grade; its length generally decreases with increasing strength. Cold-worked and high-carbon steels, Fig. 2.9 (b), exhibit a smooth transition from the initial elastic phase to the strain-hardening branch, without a distinct yield point. The yield stress of steels lacking a well-defined yield plateau is often defined as the stress at which a permanent strain of 0.2% remains after unloading, Fig. 2.9 (b); alternatively, the yield strain ε_{sy} can directly be specified. The modulus of elasticity, E_s, is roughly equal to 205 GPa for all types of steel, while yield stresses typically amount to 400...600 MPa. Unloading at any point of the stress-strain diagram occurs with approximately the same stiffness as initial loading. The elongation in the strain-hardening range occurs at constant volume (Poisson's ratio $\nu = 0.5$), resulting in a progressive reduction of the cross-sectional area. Steel stresses, in particular the tensile strength f_{su}, are usually based on the initial nominal cross-section; the actual stresses acting on the reduced area at the ultimate state may be considerably higher.

In the present work, a bilinear idealisation, Chapter 2.1, of the stress-strain response of reinforcement will frequently be applied. Using the notation of Fig. 2.9 (c), the strain-hardening modulus E_{sh} is given by

$$E_{sh} = \frac{f_{su} - f_{sy}}{\varepsilon_{su} - \varepsilon_{sy}} \qquad (2.9)$$

where $\varepsilon_{sy} = f_{sy}/E_s$ = yield strain and ε_{su} = rupture strain of reinforcement. The rupture strain ε_{su} and the ratio of tensile to yield strength, f_{su}/f_{sy}, are measures of the ductility

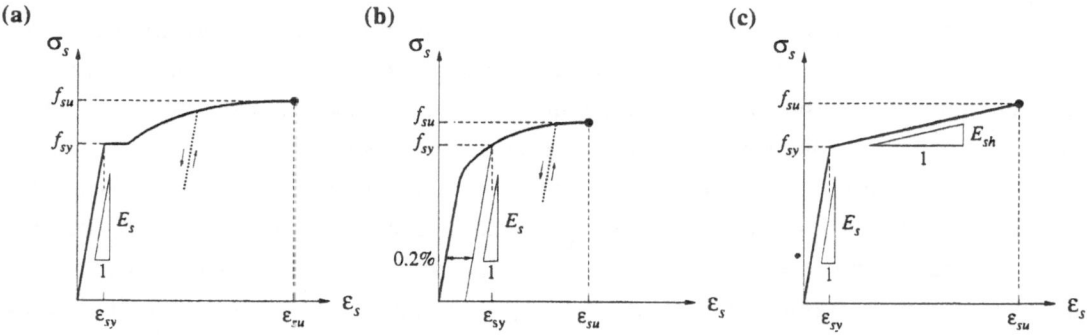

Fig. 2.9 – Stress-strain characteristics of reinforcement in uniaxial tension: (a) hot-rolled, heat-treated, low-carbon or micro-alloyed steel; (b) cold-worked or high-carbon steel; (c) bilinear idealisation.

of the steel. Hot-rolled, low-carbon or micro-alloyed steel exhibiting a stress-strain characteristic as shown in Fig. 2.9 (a) typically has higher ratios of f_{su}/f_{sy} and considerably larger rupture strains ε_{su} than cold-worked or high-carbon steel, Fig. 2.9 (b).

2.3.3 Prestressing Steel

Prestressing steel is usually cold-drawn after a homogenisation process and thus exhibits a stress-strain relationship similar to that of cold-worked reinforcement, Fig. 2.9 (b). The bilinear idealisation shown in Fig. 2.9 (c) will also be used for prestressing steel (substituting the subscript s, for reinforcing steel, by a subscript p for prestressing). Typically, yield and tensile strengths of prestressing steel are 2...3 times higher than those of ordinary reinforcing steel. The use of high-strength steel is essential to prestressing; the reinforcement strains at prestressing must be significantly higher than the long-term deformations of concrete and steel because otherwise, much of the initially applied prestress will be lost with time. On the other hand, high-strength reinforcement should not be used without prestressing since large crack widths would result from high reinforcement strains, at least in normal-strength concrete. The stiffness of seven-wire strands is lower than that of individual wires due to lateral contraction upon tensioning; typically, $E_p = 205$ GPa for wires as compared to $E_p = 195$ GPa for seven-wire strands.

2.4 Interaction of Concrete and Reinforcement

2.4.1 Bond

If relative displacements of concrete and reinforcement occur, bond stresses develop at the steel-concrete interface. The relative displacement or slip δ is given by $\delta = u_s - u_c$, where u_s and u_c denote the displacements of reinforcement and concrete, respectively. The magnitude of the bond stresses depends on the slip δ as well as on several other factors, including bar roughness (size, shape and spacing of ribs), concrete strength, position and orientation of the bar during casting, concrete cover, boundary conditions, and state of stress in concrete and reinforcement. Bond stresses are essential to the anchorage of straight rebars, they influence crack spacings and crack widths and are important if deformations of structural concrete members have to be assessed. A detailed investigation of bond and tension stiffening, including prestressed reinforcement and deformations in the plastic range of the steel stresses, can be found in a recent report by Alvarez [6].

Bond action is primarily due to interlocking of the ribs of profiled reinforcing bars and the surrounding concrete; stresses caused by adherence (plain bars) are lower by an order of magnitude. Forces are primarily transferred to the surrounding concrete by inclined compressive forces radiating out from the bars. The radial components of these inclined compressive forces are balanced by circumferential tensile stresses in the concrete or by lateral confining stresses. If significant forces have to be transmitted over a short embedment length by bond, splitting failures along the reinforcement will occur unless sufficient concrete cover or adequate circumferential reinforcement is provided; this effect is called tension splitting.

In a simplified approach, the complex mechanism of force transfer between concrete and reinforcement is substituted by a nominal bond shear stress uniformly distributed over the nominal perimeter of the reinforcing bar. Bond shear stress-slip relationships, Fig. 2.10 (b), are normally obtained from pull-out tests as shown in Fig. 2.10 (a). The average bond shear stress along the embedment length l_b can be determined from the pull-out force as

$$\tau_b = \frac{F}{\varnothing \pi l_b} \qquad (2.10)$$

where $\varnothing$ = nominal diameter of reinforcing bar. In a pull-out test, bond shear stresses increase with the slip until the maximum bond shear stress τ_{bmax} (bond strength) is reached, typically at a slip $\delta = 0.5 \ldots 1$ mm; if the slip is further increased, bond shear stresses decrease, Fig. 2.10 (b). Equilibrium requires that for any section of a structural concrete element loaded in uniform tension, Fig. 2.10 (c),

$$N = A_s \sigma_s + A_c \sigma_c, \qquad \frac{N}{A_s} = \sigma_s + \frac{(1-\rho)}{\rho} \sigma_c \qquad (2.11)$$

where $\rho = A_s/A_c$ = geometrical reinforcement ratio, A_s = cross-sectional area of reinforcement and A_c = gross cross-section of concrete. Formulating equilibrium of a differential element of length dx, Fig. 2.10 (c), one obtains the expression

$$\frac{d\sigma_s}{dx} = \frac{4\tau_b}{\varnothing}, \qquad \frac{d\sigma_c}{dx} = -\frac{4\tau_b}{\varnothing} \frac{\rho}{(1-\rho)} \qquad (2.12)$$

for the stresses transferred between concrete and reinforcement by bond. Furthermore, the kinematic condition

$$\frac{d\delta}{dx} = \frac{d}{dx}[u_s - u_c] = \varepsilon_s - \varepsilon_c \qquad (2.13)$$

is obtained from Fig. 2.10 (c) if plane sections are assumed to remain plane. Differentiating Eq. (2.13) with respect to x, inserting Eq. (2.12) and substituting stress-strain relationships for steel and concrete, a second order differential equation for the slip δ is obtained. Generally, the differential equation has to be solved in an iterative numerical manner. For linear elastic behaviour, $\sigma_s = E_s \varepsilon_s$ and $\sigma_c = E_c \varepsilon_c$, one gets

$$\frac{d^2\delta}{dx^2} = \frac{4\tau_b}{\varnothing E_s}\left(1 + \frac{n\rho}{1-\rho}\right) \qquad (2.14)$$

where $n = E_s/E_c$ = modular ratio; Eq. (2.14) can be solved analytically for certain bond shear stress-slip relationships.

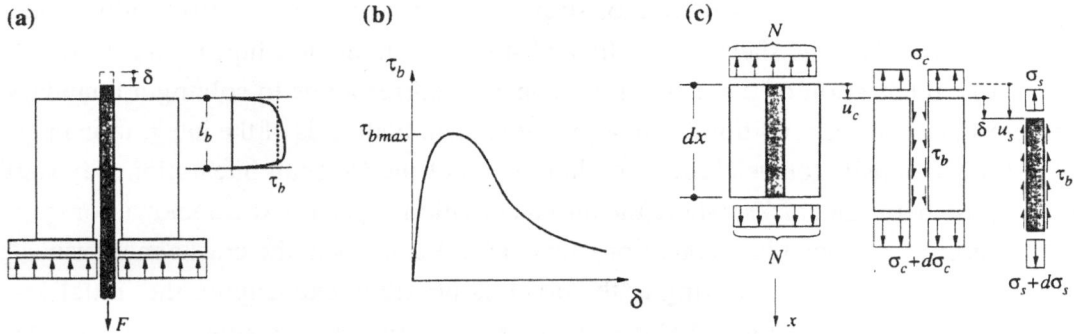

Fig. 2.10 – Bond behaviour: (a) pull-out test; (b) bond shear stress-slip relationship; (c) differential element.

2.4.2 Tension Stiffening

The effect of bond on the behaviour of structural concrete members loaded in tension is called tension stiffening, since after cracking the overall response of a structural concrete tension chord is stiffer than that of a naked steel bar of equal resistance.

The behaviour of a structural concrete tension chord can be described by a chord element bounded by two consecutive cracks, Fig. 2.11 (a). The distribution of stresses and strains within the chord element is shown in Fig. 2.11 (b) for the symmetrical case, i.e., equal tensile forces N acting on both sides of the element. At the cracks, concrete stresses are zero and the entire tensile force is carried by the reinforcement, $\sigma_{sr} = N/A_s$. Away from the cracks, tensile stresses are transferred from the reinforcement to the surrounding concrete by bond shear stresses according to Eq. (2.12). In the symmetrical case, bond shear stresses and slip vanish at the centre between cracks; there, reinforcement stresses are minimal, and the concrete stresses reach their maximum value. For a given

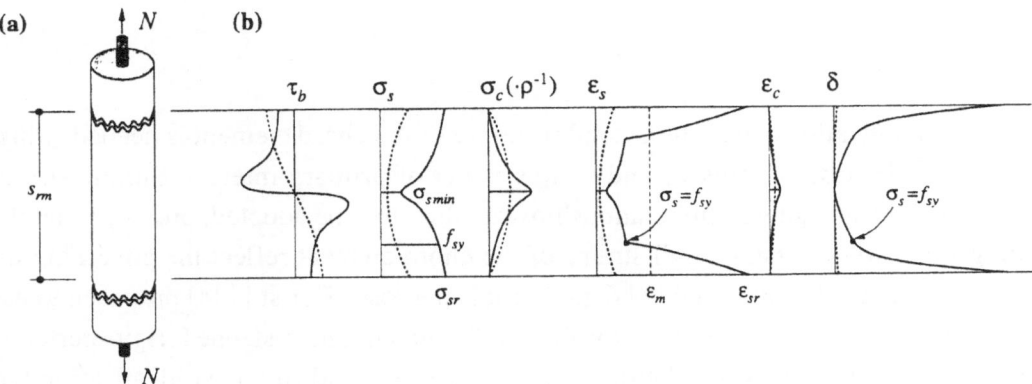

Fig. 2.11 – Tension stiffening: (a) chord element; (b) qualitative distribution of bond shear stresses, steel and concrete stresses and strains, and bond slip.

applied tensile force, the distribution of stresses and strains, Fig. 2.11 (b), can be determined for arbitrary bond shear stress-slip and stress-strain relationships from Eqs. (2.12) and (2.13). Integration of the differential equation corresponds to solving a boundary problem since certain conditions have to be satisfied at both ends of the integration interval. For equal tensile forces N acting on both sides of the element, integration may start at the centre between cracks, where the initial conditions $u_s = u_c = 0$ are known for symmetry reasons; as a boundary condition, the concrete stresses at the cracks must vanish. Alternatively, integration starting at the crack is possible, exchanging the initial and boundary conditions mentioned above. If the tensile force varies along the chord element, the section at which $u_s = u_c = 0$ is not known beforehand and the solution is more complicated; suitable algorithms and a detailed examination are given in a recent report by Alvarez [6]. Apart from a general discussion of tension stiffening effects in the web of concrete girders presented in Chapter 6.2.3, only the symmetrical case with equal tensile forces N acting on both sides of the element will be applied in this thesis.

Observing that the concrete tensile stresses cannot be greater than the concrete tensile strength f_{ct}, one obtains the requirement

$$\frac{4}{\varnothing} \frac{\rho}{(1-\rho)} \int_{x=0}^{s_{rmo}/2} \tau_b dx \le f_{ct} \tag{2.15}$$

for the maximum crack spacing s_{rmo} in a fully developed crack pattern. The minimum crack spacing amounts to $s_{rmo}/2$ since a tensile stress equal to the concrete tensile strength must be transferred to the concrete in order to generate a new crack [144,93]. Hence, the crack spacing s_{rm} in a fully developed crack pattern is limited by

$$\frac{s_{rmo}}{2} \le s_{rm} \le s_{rmo} \tag{2.16}$$

or, equivalently, $0.5 \le \lambda \le 1$, where

$$\lambda = \frac{s_{rm}}{s_{rmo}} \tag{2.17}$$

For most applications, only the overall response of the chord element is needed, while the exact distribution of stresses and strains is not of primary interest. Simple stress-strain and bond shear stress-slip relationships can therefore be adopted, provided that the resulting steel stresses and overall strains of the chord element reflect the governing influences and match the experimental data. For this purpose, Sigrist [144] proposed to use a bilinear stress-strain characteristic for the reinforcement and a stepped, rigid-perfectly plastic bond shear stress-slip relationship, Figs. 2.12 (a) and (b). This idealisation has been called "tension chord model" [144,7,93,6,94]. For the bond shear stresses prior to and after the onset of yielding of the reinforcement, $\tau_{bo} = 2f_{ct}$ and $\tau_{b1} = f_{ct}$ is assumed, respectively, where f_{ct} = tensile strength of concrete, see Chapter 2.2.1.

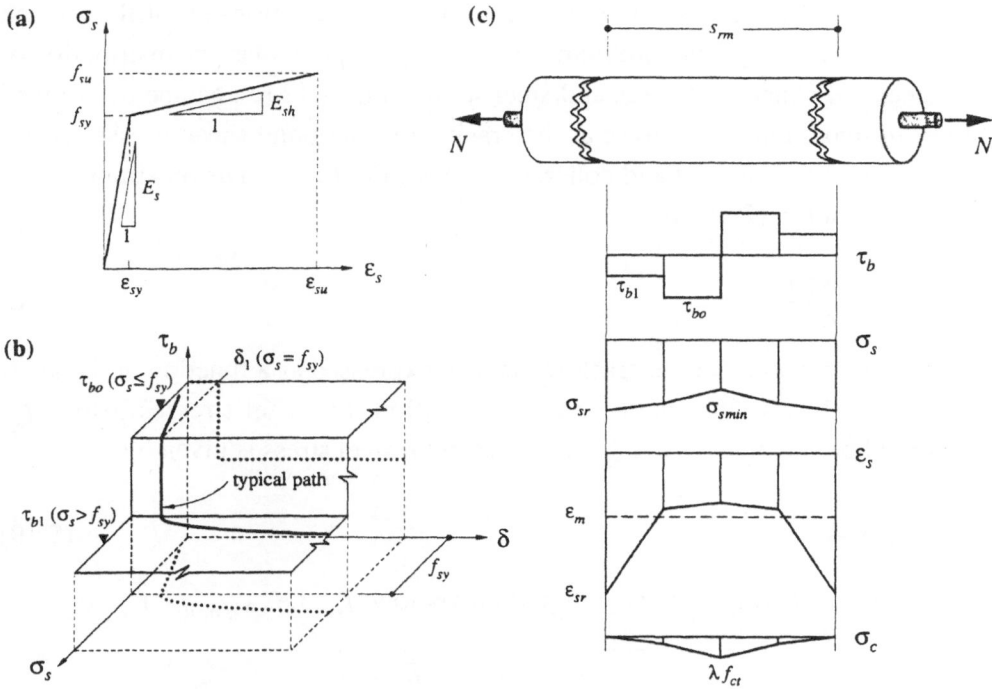

Fig. 2.12 – Tension chord model: (a) stress-strain diagram for reinforcement; (b) bond shear stress-slip relationship; (c) chord element and distribution of bond shear, steel and concrete stresses, and steel strains.

The sudden drop of bond shear stresses at the onset of yielding, $\delta = \delta_1$, Fig. 2.12 (b), seems somewhat arbitrary. However, a closer examination of the underlying phenomena reveals that a stepped, perfectly plastic idealisation is indeed appropriate. The response of the reinforcement, Fig. 2.12 (a), is characterised by a sudden change in the stress-strain curve at $\sigma_s = f_{sy}$, whereafter steel strains increase at a much faster rate; typical hardening moduli of steel are about 100 times lower than E_s. Consequently, substantially larger strains and slips occur after the onset of yielding than in the elastic range of steel stresses, Fig. 2.11 (b), resulting in significantly lower bond shear stresses, Fig. 2.10 (b). Due to the rapid growth of steel strains in the strain-hardening range, only a small portion of the chord element undergoes slip immediately to the right of δ_1, and a smooth reduction of bond shear stresses after the onset of yielding would only slightly alter the overall behaviour. Furthermore, large steel strains and slips contribute to a progressive deterioration of bond near the cracks, and cross-sectional areas of the bars are gradually reduced (Chapter 2.3.2), especially in the strain-hardening range; when longitudinal strains become large, the reduction in diameter of the bars will contribute to further deterioration of bond. While actual bond shear stress-slip relationships observed from tests are much more complicated and more sophisticated idealisations are certainly possible, the proposed stepped, perfectly plastic bond shear stress-slip relationship represents the simplest possible formulation capable of reflecting the reduction of bond shear stresses after the onset of yielding observed in experiments [141,144,94].

The stepped rigid-perfectly plastic bond shear stress-slip relationship of the tension chord model allows treating many problems analytically. In particular, the distribution of bond shear, steel and concrete stresses and steel strains, Fig. 2.12 (c), can be determined for any assumed maximum steel stress at the crack; constant bond shear stresses correspond to linear variations of steel and concrete stresses, Eq. (2.12). The maximum crack spacing follows from Eq. (2.15) as

$$s_{rmo} = \frac{f_{ct} \varnothing}{2\tau_{bo}} \frac{(1-\rho)}{\rho} \tag{2.18}$$

and the maximum steel stress at the crack σ_{sr} can be expressed as a function of the average strain, ε_m, which describes the overall deformation. For steel stresses below f_{sy} along the entire chord element, $\sigma_{sr} \le f_{sy}$, the maximum steel stress is given by

$$\sigma_{sr} = E_s \varepsilon_m + \frac{\tau_{bo} s_{rm}}{\varnothing} \tag{2.19}$$

while for steel stresses partially above and partially below f_{sy}, i.e., $\sigma_{s\,min} \le f_{sy} < \sigma_{sr}$

$$\sigma_{sr} = f_{sy} + 2 \frac{\dfrac{\tau_{bo} s_{rm}}{\varnothing} - \sqrt{(f_{sy} - E_s \varepsilon_m) \dfrac{\tau_{b1} s_{rm}}{\varnothing} \left(\dfrac{\tau_{bo}}{\tau_{b1}} - \dfrac{E_s}{E_{sh}}\right) + \dfrac{E_s}{E_{sh}} \tau_{bo} \tau_{b1} \dfrac{s_{rm}^2}{\varnothing^2}}}{\dfrac{\tau_{bo}}{\tau_{b1}} - \dfrac{E_s}{E_{sh}}} \tag{2.20}$$

and for steel stresses above f_{sy} along the entire chord element, i.e., $f_{sy} < \sigma_{s\,min}$

$$\sigma_{sr} = f_{sy} + E_{sh}\left(\varepsilon_m - \frac{f_{sy}}{E_s}\right) + \frac{\tau_{b1} s_{rm}}{\varnothing} \tag{2.21}$$

Fig. 2.13 (a) illustrates the above equations for three different reinforcement ratios, assuming $\lambda = 1$, $f_{sy} = 500$ MPa, $f_{su} = 625$ MPa, $E_s = 200$ GPa, $\varepsilon_{su} = 0.05$, $\varnothing = 16$ mm, $f_c' = 30$ MPa and $\tau_{b1} = \tau_{bo}/2 = f_{ct}$.

The response of cracked reinforced concrete members in tension is conventionally expressed in terms of maximum steel stresses at cracks and average strains of the member, Eqs. (2.19), (2.20) and (2.21), since maximum steel stresses govern failure while average deformations are important for serviceability calculations. Empirical relationships combining average strains with average (over the length of the chord element) steel and concrete tensile stresses, σ_{sm} and σ_{cm}, have also been proposed [57,156], primarily in order to describe tension stiffening effects in structural concrete panels loaded in plane stress (Chapter 4.3.3). Though of little physical significance, average stress-average strain relationships resulting from the tension chord model are established below, primarily for comparison purposes with the empirical relationships proposed in [57,156].

If steel stresses are either below or above the yield stress over the entire length of the chord element, the average stress-average strain curve of steel matches the stress-strain curve of naked steel, i.e., $\sigma_{sm} = E_s \varepsilon_m$ and $\sigma_{sm} = f_{sy} + E_{sh}(\varepsilon_m - f_{sy}/E_s)$, while average concrete tensile stresses are constant at

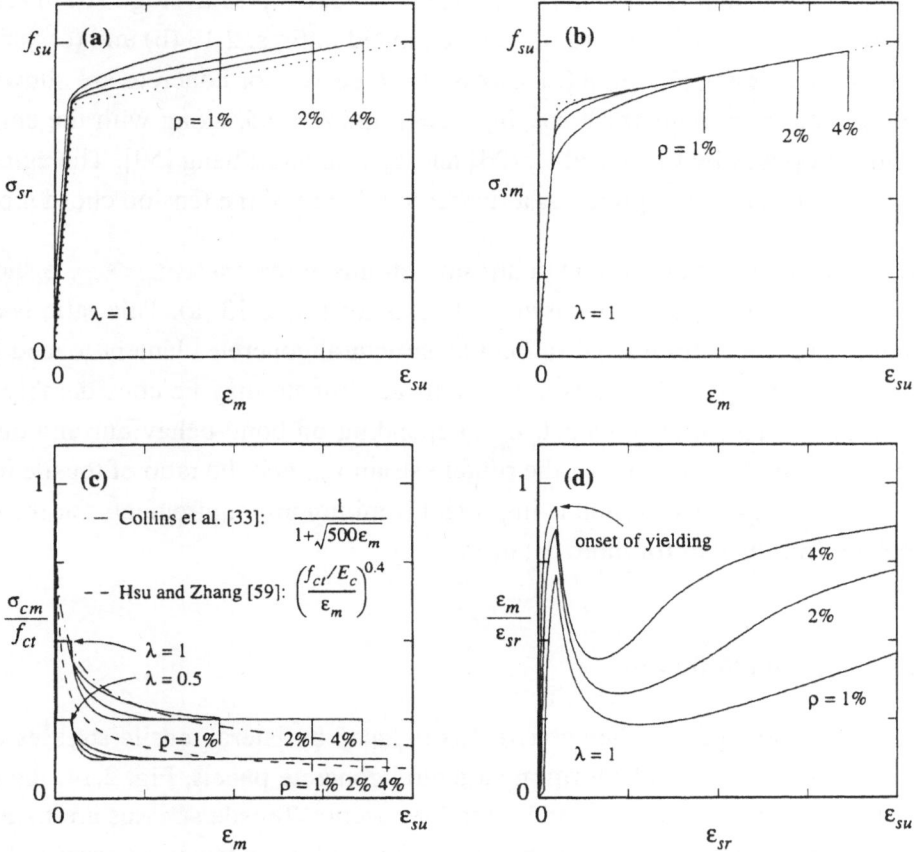

Fig. 2.13 – Tension chord model: (a) maximum steel stresses, (b) average steel stresses, (c) average concrete tensile stresses versus average strains; (d) strain localisation versus steel strains at cracks.

$$\sigma_{cm} = \frac{\tau_{bo}s_{rm}}{\varnothing}\frac{\rho}{(1-\rho)} \quad \text{and} \quad \sigma_{cm} = \frac{\tau_{b1}s_{rm}}{\varnothing}\frac{\rho}{(1-\rho)} \tag{2.22}$$

respectively. For steel stresses partially above and partially below the yield stress, average steel stresses can be expressed as

$$\sigma_{sm} = f_{sy} - \frac{(\sigma_{sr}-f_{sy})^2\varnothing}{4\tau_{b1}s_{rm}}\left(\frac{\tau_{bo}}{\tau_{b1}}-1\right) + (\sigma_{sr}-f_{sy})\frac{\tau_{bo}}{\tau_{b1}} - \frac{\tau_{bo}s_{rm}}{\varnothing} \tag{2.23}$$

with σ_{sr} according to Eq. (2.20). Since equilibrium must be fulfilled at every section over the entire length of the chord element, Eq. (2.11), average steel and concrete tensile stresses are related by

$$\sigma_{sm} + \frac{(1-\rho)}{\rho}\sigma_{cm} = \sigma_{sr} \tag{2.24}$$

from which the average concrete stresses can be calculated. Eq. (2.24) can also be used to verify empirical average stress-average strain relationships established separately for reinforcement and concrete (Chapter 4.3.3). The average stress-average strain relationships according to the tension chord model are plotted in Figs. 2.13 (b) and (c), using the same assumptions as for Fig. 2.13 (a). Stress-strain curves for concrete are shown both for maximum and minimum crack spacing, $\lambda = 1$ and $\lambda = 0.5$, along with the empirical relationships proposed by Collins et al. [33] and by Hsu and Zhang [59]. The agreement is quite good and can be interpreted as a further validation of the tension chord model.

The ratio of overall strains to maximum steel strains at the crack, $\varepsilon_m/\varepsilon_{sr}$, is shown in Fig. 2.13 (d), again using the same assumptions as for Fig. 2.13 (a). This ratio is an important measure for the strain localisation in a structural concrete element loaded in tension. As seen from Fig. 2.13 (d), overall strains at ultimate may be considerably lower than the rupture strain of naked steel, ε_{su}, depending on bond behaviour and ductility properties of the reinforcement, i.e., the rupture strain ε_{su} and the ratio of tensile to yield strength, f_{su}/f_{sy}. This observation is important if minimum requirements for the ductility of steel are formulated; for more details see [6].

2.4.3 Compression Softening

The compressive strength of plain concrete is reduced by lateral tensile stresses as outlined in Chapter 2.2.3. If cracks form in structural concrete panels, Fig. 2.14, the tensile force is taken by the reinforcement and no failure occurs. Tensile stresses are transferred to the concrete between the cracks by bond stresses, and these tensile stresses will eventually promote laminar splitting failures at compressive stresses somewhat lower than the uniaxial compressive strength f_c, Chapter 2.2.2. On the other hand, the concrete is laterally constrained by non-yielding reinforcement, resulting in slightly enhanced compressive strength and ductility. However, this effect is much less pronounced than for triaxial confinement, Chapter 2.4.4, because the deformations of the panel normal to its plane are unrestricted. In summary, the compressive strength of structural concrete panels will be similar to that of plain concrete as long as the reinforcement does not yield.

If the concrete has been deteriorated as a result of large transverse strains, the compressive strength will be further reduced. This effect is called compression softening; reductions of up to 80% as compared to f_c' have been reported [155]. The drastic reduction of the compressive strength of concrete at large lateral strains cannot be attributed to large crack widths; the strength of the concrete struts between the cracks is only moderately reduced by the crack opening. Rather, as the slip between concrete and reinforcement increases, the ribs of the reinforcing bars tend to separate the surrounding concrete along the reinforcement (tension splitting, Chapter 2.4.1), and laminar splitting of the concrete will occur at markedly lower compressive stresses than in uniaxial compression. While specimens with a single reinforcement layer will inevitably fail, spalling of the concrete cover will occur for two or more reinforcement layers, and a further load increase is often possible. Strictly speaking, the compression softening of concrete is thus not a material property, since it depends on the type and the layout of the reinforcement.

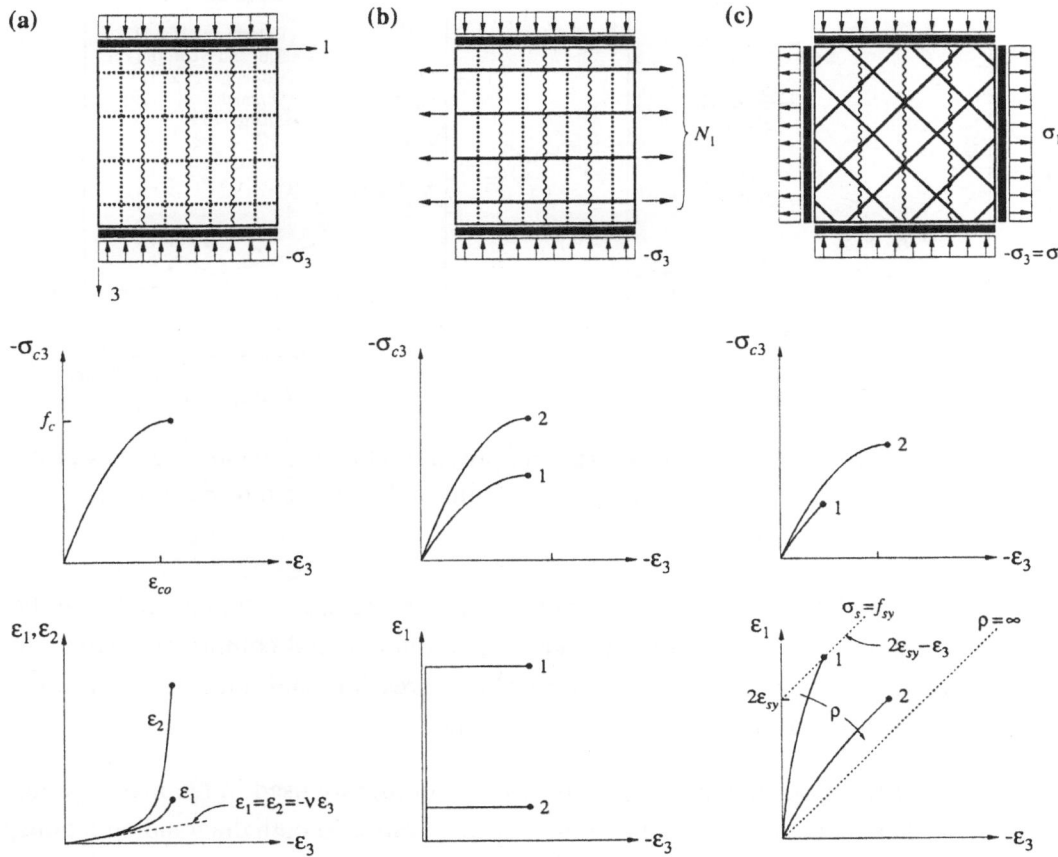

Fig. 2.14 – Compression softening tests: (a) uniaxial compression; (b) sequential tension-compression; (c) shear test (isotropically reinforced panel).

Different testing methods have been used in order to investigate the compression softening of concrete. Fig. 2.14 compares stresses and strains in such tests with those in a uniaxially compressed panel whose response, Fig. 2.14 (a), closely matches that of an axisymmetrical specimen, Figs. 2.5 (a) and (d). Reinforcement in the compressive direction directly resists compression and enhances the ductility of the panel by balancing local variations of concrete properties. Lateral reinforcement results in slightly improved strength and ductility as already mentioned. The typically observed laminar splitting failure is reflected by the stronger increase of the lateral strain normal to the stress-free plane, ε_2, as compared to the lateral strain in the plane of the panel, ε_1. Fig. 2.14 (b) shows a panel with reinforcement in the tensile direction subjected to sequential loading in tension-compression. With this type of test, arbitrary transverse strains, $\varepsilon_1 < \varepsilon_{su}$, can be investigated. In a first phase of the test, lateral tension N_1 is applied until a desired transverse strain ε_1 is obtained; the panel displays a regular crack pattern at this stage. Subsequently, compressive stresses are increased at a constant lateral strain ε_1 until the panel fails in compression. The compressive strength obtained in this way is lower than the strength in uniaxial compression if ε_1 is kept constant at a higher value than ε_1 at the ultimate state in uniaxial compression. For smaller values of ε_1, compressive forces have

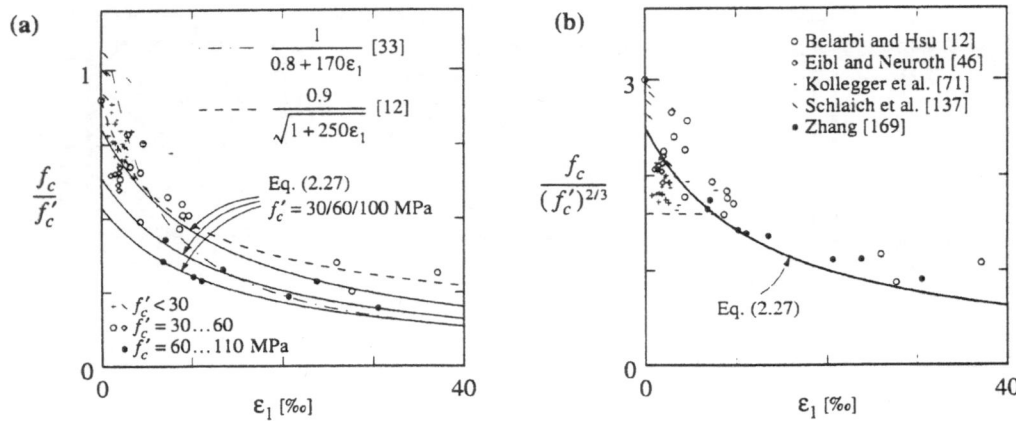

Fig. 2.15 – Tension-compression tests: (a) compressive strength versus lateral strains; (b) ratio $f_c/(f_c')^{2/3}$ versus lateral strains and compared to Eq. (2.27).

to be applied in the lateral direction as the compressive strength is approached, and the strength of the panel will be equal to or greater than the uniaxial compressive strength. The failure mode corresponds to that in uniaxial compression, and reinforcement in the compressive direction has the same beneficial effects.

Specimens of the type shown in Fig. 2.14 (c) are commonly used in the investigation of membrane shear behaviour. For equal reinforcement ratios in both diagonal directions, stresses in the reinforcement and in the concrete at the cracks can be determined from simple considerations. However, concrete compressive stresses between the cracks are smaller than at the cracks, see Chapter 5. Still, such tests may be used in the investigation of compression softening. Tension-compression tests typically yield slightly lower compressive strengths than for reinforcement in the tensile direction; the inclination of the principal compressive direction with respect to the reinforcement thus affects the compression softening behaviour. In shear tests, the load is applied such that $\sigma_3 = -\sigma_1$, corresponding to concrete compressive stresses $\sigma_{c3} = 2\sigma_3$ and steel stresses $\sigma_s = \sigma_1/\rho$ in both reinforcements (Chapter 3.3). As the load is increased, the lateral strain ε_1 grows faster than the compressive strain ε_3, depending on the reinforcement ratio, Fig. 2.14 (c). Eventually, the panel will fail by yielding of the reinforcement or crushing of the concrete. In the latter case, the reinforcement is still elastic and will enhance both the strength and the ductility of the concrete. From Fig. 2.14 (c) it is clear that only a limited range of transverse strains, $\varepsilon_1 < 2\varepsilon_{sy}+\varepsilon_{co}$, can be investigated in shear tests of isotropically reinforced panels. Shear tests of orthotropically reinforced panels with large transverse strains have also been used to calibrate compression softening relationships. In such tests, concrete and reinforcement stresses are determined based on several assumptions, see Chapter 5. Thus, to a certain degree, the test results reflect the theory used in assessing the stresses.

For the reasons outlined above, only tension-compression tests with reinforcement in the tensile direction are considered here. Tests of orthogonally reinforced panels loaded in shear are examined in Chapter 5 and Appendix B.

Robinson and Demorieux [130] were the first researchers who explicitly investigated the dependence of the compressive strength of concrete on transverse strains. Based on numerous tests, Collins [28] and Vecchio and Collins [154,155] proposed various expressions, leading to the equation [20,33]

$$f_c = \frac{f_c'}{0.8 + 170\varepsilon_1} \le f_c' \tag{2.25}$$

for the peak compressive stress f_c as a function of ε_1. Note that ε_1 is the average lateral strain, Chapter 2.4.2, not the strain of the concrete between cracks. Fig. 2.15 (a) compares Eq. (2.25) with test results and with the compression softening relationship proposed by Hsu [57] and Belarbi and Hsu [12]. The test results shown in Fig. 2.15 support the observation that the compressive strength f_c increases less than proportional with f_c', Chapter 2.2.2. Muttoni [108] proposed a value for the compressive strength under moderate transverse strains, i.e.,

$$f_c = 1.6(f_c')^{2/3} \quad \text{in MPa, where} \quad f_c \le 0.6 f_c' \tag{2.26}$$

including the influence of f_c', but neglecting the detrimental influence of lateral strains. Based on the evaluation of available test results of tension-compression tests with reinforcement in the tensile direction, the following expression for the concrete compressive strength is proposed

$$f_c = \frac{(f_c')^{2/3}}{0.4 + 30\varepsilon_1} \quad \text{in MPa, where} \quad f_c \le f_c' \tag{2.27}$$

This expression reflects the influence of f_c' as well as that of ε_1. Eq. (2.27) is plotted in Figs. 2.15 (a) and (b); the correlation with experimental evidence is satisfactory. The validation of Eq. (2.27) for the prediction of the compressive strength of concrete in membrane elements is given in Chapter 5 and in Appendix B.

A recent survey of research on the compression softening of concrete [131] revealed that several researchers suggest a much smaller reduction of the compressive strength than observed in the tests underlying Eq. (2.27). These researchers conclude that the reduction of the compressive strength is no more than 20% as compared to f_c'. In their tests [46,71,137], they investigated only limited lateral strains of about $\varepsilon_1 < 2\varepsilon_{sy}$ and consequently, they obtained only small reductions of the compressive strength, see Fig. 2.15 (a). They argue that higher transverse strains do not have to be taken into account because the ultimate load of a structural concrete member is attained at strains of $\varepsilon_1 > \varepsilon_{sy}$, i.e., concrete never undergoes higher transverse strains. However, in orthotropically reinforced membrane elements and thin-webbed beams with low stirrup reinforcement ratios, the load can be substantially increased after yielding of the weaker reinforcement, and transverse strains at the ultimate state are often significantly greater than ε_{sy}, Chapters 5 and 6. Reductions of the compressive strength in the range predicted by Eq. (2.27) have to be expected in such situations.

Material Properties

2.4.4 Confinement

Both the strength and the ductility of concrete are significantly enhanced by lateral compressive stresses (Chapter 2.2.4). Instead of applying a lateral pressure, a similar effect can be obtained by providing closely spaced reinforcement in the form of ties or spirals confining the core concrete. Mörsch [105] pointed out that spirals in columns are more effective than longitudinal bars, i.e., the ultimate resistance of a column will be greater if the same amount of steel is used as spiral reinforcement rather than as longitudinal reinforcement directly resisting axial forces.

If a short concrete column reinforced by closely spaced hoops is loaded in compression, the concrete tends to expand as the uniaxial compressive strength is approached, Chapter 2.2.2, causing the hoops to exert confinement on the concrete. The resulting lateral compressive stresses increase the strength of the confined concrete, similar to actively applied lateral pressure. Generally, the assumption is made that lateral reinforcement yields at the ultimate state, from which confining stresses can easily be calculated. Some tests on spirally reinforced high-strength concrete columns [96] exhibited lateral strains at the ultimate state below the yield strain of the spirals, indicating that existing design approaches should be cautiously used for high-strength columns. The concrete in these columns was made from mixtures with very high cement content and low water-cement ratios, but without fly-ash or microsilica additives. Therefore, recent test results of high-strength concrete columns [56,133,134,166] will be compared here with the well-known tests by Richart et al. [128] and Somes [145] as well as existing design approaches.

In circular columns, Fig. 2.16 (a), the confining pressures are uniformly distributed over the perimeter, while a non-uniform distribution exists in the axial direction. Based on a stress-field approach (Chapter 3.3.3) for the confining stresses, Kanellopoulos [61] proposed the following expression for the strength of the confined concrete inside the spirals or hoops

$$\frac{f_{c3}}{f_c} = 1 + 4\omega\left(1 - \frac{s}{d}\right)^2\left(1 - \frac{\pi s}{4d}\right)^{-1} \quad \text{where} \quad \omega = \frac{A_s f_{sy}}{sd f_c} \quad (2.28)$$

and A_s is the cross-sectional area of all reinforcement per height s crossing a plane containing the column axis (i.e., twice the cross-sectional area of a conventional hoop).

Sigrist [144] presented a comprehensive examination of the behaviour of rectangular columns. Based on a stress-field approach for the confining pressure, he proposed an expression for the strength of the confined concrete inside the hoops, which, for a square column with equal reinforcement in both directions, Fig. 2.17 (a), simplifies to

$$\frac{f_{c3}}{f_c} = 1 + 4\omega\left(1 - \frac{s}{a}\right)^2 \quad \text{where} \quad \omega = \frac{A_s f_{sy}}{sa f_c} \quad (2.29)$$

and A_s is the cross-sectional area of all reinforcement per height s crossing a plane parallel to a side face of the column and containing the column axis (i.e., twice the cross-sec-

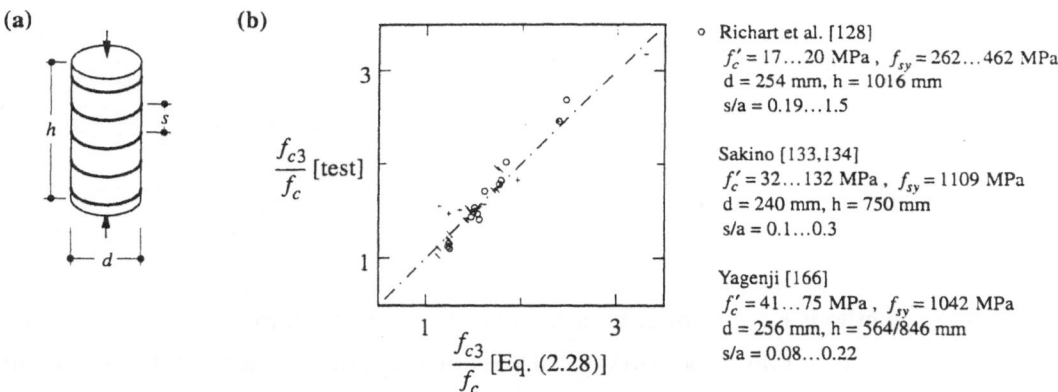

(a)

(b)

$\frac{f_{c3}}{f_c}$ [test]

Richart et al. [128]
$f_c' = 17...20$ MPa , $f_{sy} = 262...462$ MPa
d = 254 mm, h = 1016 mm
s/a = 0.19...1.5

Sakino [133,134]
$f_c' = 32...132$ MPa , $f_{sy} = 1109$ MPa
d = 240 mm, h = 750 mm
s/a = 0.1...0.3

Yagenji [166]
$f_c' = 41...75$ MPa , $f_{sy} = 1042$ MPa
d = 256 mm, h = 564/846 mm
s/a = 0.08...0.22

$\frac{f_{c3}}{f_c}$ [Eq. (2.28)]

Fig. 2.16 – Tests on short circular columns: (a) notation; (b) comparison of theoretical and experimental resistances.

tional area of a conventional hoop). Note that the concrete strengths are normalised with respect to that of an unconfined specimen, f_c, rather than f_c', cf. Chapter 2.2.2.

Figs. 2.16 (b) and 2.17 (b) compare results of tests on normal and high-strength concrete columns with the predictions according to Eqs. (2.28) and (2.29). The agreement is satisfactory, with no significant differences between normal and high-strength concrete specimens. Considering the fact that the high-strength concrete specimens contained hoops made from high-strength reinforcement with correspondingly high yield strains, the present test results indicate that lateral expansion in high-strength concrete columns is high enough to activate the full yield stress of the confining reinforcement. The high-strength columns analysed here were made from high-strength concrete with fly-ash or microsilica additives in contrast to the earlier tests which used very high cement contents; this may explain the different behaviour of the specimens. In summary, commonly used design approaches have been validated for the application to high-strength concrete columns by the test results presented here.

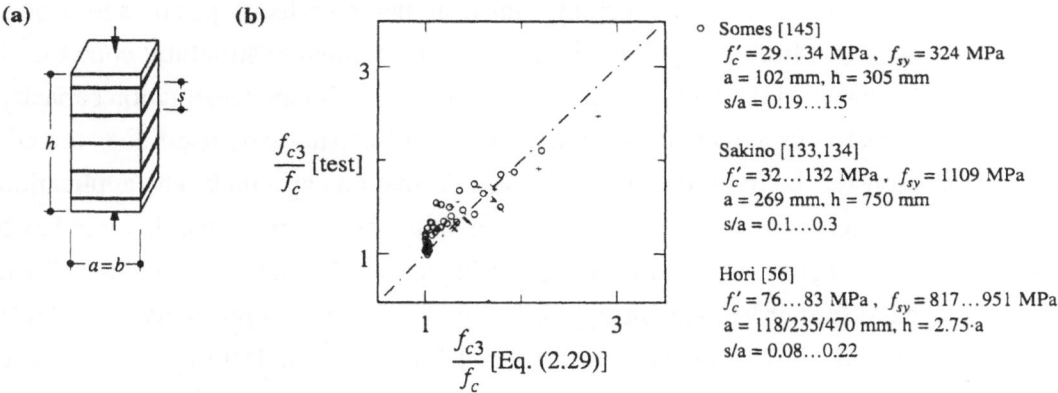

(a)

(b)

$\frac{f_{c3}}{f_c}$ [test]

Somes [145]
$f_c' = 29...34$ MPa , $f_{sy} = 324$ MPa
a = 102 mm, h = 305 mm
s/a = 0.19...1.5

Sakino [133,134]
$f_c' = 32...132$ MPa , $f_{sy} = 1109$ MPa
a = 269 mm, h = 750 mm
s/a = 0.1...0.3

Hori [56]
$f_c' = 76...83$ MPa , $f_{sy} = 817...951$ MPa
a = 118/235/470 mm, h = 2.75·a
s/a = 0.08...0.22

$\frac{f_{c3}}{f_c}$ [Eq. (2.29)]

Fig. 2.17 – Tests on short square columns: (a) notation; (b) comparison of theoretical and experimental resistances.

3 Limit Analysis of Structural Concrete

3.1 General

Limit analysis methods have implicitly or explicitly been applied to the solution of engineering problems for a long time; surveys of the corresponding historical developments were given by Prager [123] and Chen [24]. After having been thrust into the background for many decades by the emerging theory of elasticity, limit analysis methods were put on a sound physical basis around 1950 through the development of the theory of plasticity and have recently regained the attention of engineers.

As pointed out by Melan [101], it is virtually impossible to determine the actual state of stress in a structure since significant but practically unknown self-equilibrated (or residual) states of stress occur in any structure. Structural analyses based on the theory of elasticity (which typically start from the assumption of zero residual stresses) are therefore inherently debatable. However, ultimate loads of sufficiently ductile structures are independent of residual stresses and hence, limit analysis methods provide a suitable basis for strength design. Even for complex problems a realistic estimate of the ultimate load can be obtained with relatively little computational effort. Often, closed form solutions for the ultimate load can be derived; the resulting expressions directly reflect the influences of the governing parameters and the geometry of the problem, giving engineers a clear idea of the load carrying behaviour and of possible collapse mechanisms of the structure under consideration.

Chapter 3.2 summarises the theory of plastic potential and the basic theorems of limit analysis for perfectly plastic materials. While typically, reinforcing steel exhibits a rather ductile behaviour, the response of concrete is far from being perfectly plastic and hence, it has been argued that limit analysis methods cannot be applied to structural concrete at all. Indeed, application of the theory of plasticity requires sufficient deformation capacity of all structural members and elements, and the theory of plasticity by itself does not address the questions of the required and provided deformation capacities. The application of limit analysis methods to structural concrete is thus not straightforward; some basic aspects of the related problems are discussed in Chapter 3.3. In Chapters 5 and 6, refined models are used to further validate the applicability of the theory of plasticity to concrete structures subjected to in-plane shear and normal forces, particularly for cases where failure is governed by crushing of the concrete.

3.2 Limit Analysis of Perfect Plasticity

In the present chapter, limit analysis methods will be outlined for perfectly plastic material behaviour. Only small deformations are considered, such that geometry changes at the ultimate state are insignificant; the principle of virtual work can thus be applied to the undeformed body. Rigid-perfectly plastic material behaviour is assumed, Fig. 2.1 (f); this allows avoiding complicated formulations involving elastic strains. Since only infinitesimal deformations are considered, stresses remain constant at collapse for perfectly plastic behaviour [42,44] and hence, no elastic strain increments occur at the ultimate state if elastic-perfectly plastic material is considered, Fig. 2.1 (e). The limit theorems established for rigid-perfectly plastic material behaviour are thus also valid for elastic-perfectly plastic material.

3.2.1 Theory of Plastic Potential

Plastic strain increments $d\varepsilon^{(p)}$ or equivalently, plastic strain rates $\dot{\varepsilon}^{(p)}$ are of primary interest at collapse. A failure mechanism is determined by the ratios of the plastic strain increments; their absolute values are irrelevant since only infinitesimal deformations are considered. In order to avoid complex formulations, the notation $\dot{\varepsilon}$ without the superscript $^{(p)}$ for the plastic strain rates will be used. The strain rates $\dot{\varepsilon}$ do not represent differentiation of the strains with respect to the physical time t; rather, t is just a scalar, and products $\sigma\cdot\dot{\varepsilon}$ are thus termed work or dissipation rather than power or rate of dissipation.

Fig. 3.1 (a) shows a yield surface determined by the yield condition $\Phi = \Phi(\sigma) = 0$, enclosing an aplastic domain $\Phi(\sigma) < 0$. For states of stress below the yield limit, $\Phi(\sigma) < 0$, the body remains rigid, while for stress combinations at the yield surface, $\Phi(\sigma) = 0$, plastic flow may occur. Assuming that the aplastic domain is convex and that the plastic strain increments at failure are perpendicular to the yield surface, i.e., that the associated flow or normality rule is valid,

$$\dot{\varepsilon} = \kappa\cdot\mathrm{grad}\Phi \qquad (3.1)$$

where κ denotes an arbitrary non-negative factor, one obtains

$$(\sigma - \overset{*}{\sigma})\cdot\dot{\varepsilon} \geq 0 \qquad (3.2)$$

where σ is the actual stress state at the yield surface corresponding to $\dot{\varepsilon}$, and $\overset{*}{\sigma}$ is any other stress state at or within the yield surface, Fig. 3.1 (a). Rearranging Eq. (3.2) one obtains

$$dD(\dot{\varepsilon}) = \sigma\cdot\dot{\varepsilon} \geq \overset{*}{\sigma}\cdot\dot{\varepsilon} \qquad (3.3)$$

which is the principle of maximum energy dissipation postulated by von Mises [102]. According to this principle, the (virtual) dissipation per unit volume done on a given plastic strain rate $\dot{\varepsilon}$ assumes a maximum for the associated (or compatible) state of stress $\boldsymbol{\sigma}$. For any yield surface, the actual dissipation dD per unit volume is uniquely determined by the strain rate vector $\dot{\boldsymbol{\varepsilon}}$.

The principle of maximum dissipation, Eq. (3.3), has been derived above from the assumptions of convexity of the yield surface and normality of the plastic strain rates. Alternatively, the principle of maximum dissipation can be postulated, and convexity of the yield surface as well as the associated flow rule follow from it. Considering Eq. (3.1), the yield condition Φ has been called plastic potential, and the above relations altogether are known as the theory of plastic potential. This theory, outlined so far only for a unit volume, can be extended to entire bodies or systems. The yield condition of the system can be expressed in terms of the total stress vector S, representing the integral of all local stress vectors, as $\Phi(S) = 0$, Fig. 3.1 (b). Often, kinematic restrictions – such as Bernoulli's hypothesis of plane sections remaining plane in beam theory – are introduced, and the system is analysed in terms of generalised stresses Z and generalised strains $\dot{z}$ as proposed by Prager [122]. The kinematic restrictions correspond to a projection of the yield surface from the stress space, $\Phi(S) = 0$, to the generalised stress space, $\Phi(Z) = 0$, as illustrated in Fig. 3.1 (c) for a three-dimensional stress space and a two-dimensional generalised stress space. The stress components "lost" in the projection – such as the shear force in connection with Bernoulli's hypothesis – are termed generalised reactions since they can assume arbitrary values, as required for equilibrium, but they do not contribute to the dissipation. If all elements of the system obey the theory of plastic potential, this theory applies to generalised stresses and strains as well [171].

Representing entire load-cases rather than specifying loading histories by the distribution of surface and body forces, applied loads can be expressed by generalised forces F. Introducing associated generalised deformations $\dot{f}$ such that the work done by the generalised forces equals $W_d = F \cdot \dot{f}$, the theory of plastic potential remains valid in terms of generalised forces and deformations [122,171]. A detailed discussion of these subjects, including proofs of the extension of the theory of plastic potential to generalised stresses and strains as well as to generalised forces and deformations was given by Marti [82].

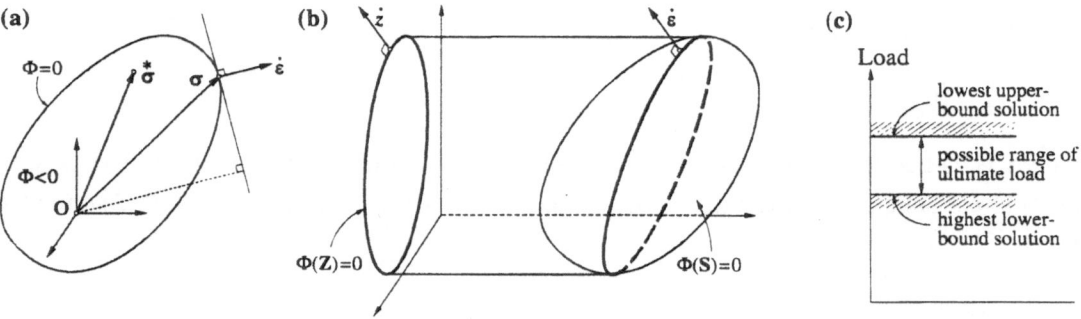

Fig. 3.1 – Limit analysis: (a) yield surface and plastic strain increment; (b) projection of yield surface from stress space to generalised stress space; (c) application of upper- and lower-bound theorem.

3.2.2 Theorems of Limit Analysis

Application of the principle of virtual work to a perfectly plastic body obeying the theory of plastic potential yields the basic theorems of limit analysis established by Gvozdev [51], Hill [52], and Drucker, Greenberg and Prager [42,44]:

- **Lower-bound theorem**: Any load corresponding to a statically admissible state of stress everywhere at or below yield is not higher than the ultimate load.

- **Upper-bound theorem**: Any load resulting from considering a kinematically admissible state of deformation and setting the work done by the external forces equal to the internal energy dissipation is not lower than the ultimate load.

For coinciding lower- and upper-bound solutions a uniqueness theorem was expressed by Drucker, Greenberg, Lee and Prager [43] and generalised by Sayir and Ziegler [136]:

- **Uniqueness theorem**: Any load for which a complete solution, i.e. a statically admissible state of stress everywhere at or below yield and a compatible, kinematically admissible state of deformation can be found, is equal to the ultimate load.

A state of stress is statically admissible if it fulfils equilibrium and static boundary conditions. A state of deformation is kinematically admissible if it fulfils kinematic relations and kinematic boundary conditions. A state of stress and a state of deformation are compatible if they obey the theory of plastic potential. In a complete solution, the states of stress and deformation merely have to be compatible in the sense stated above.

The static or lower-bound method of the theory of plasticity is based on the lower-bound theorem. Starting from statically admissible states of stress or stress fields everywhere at or below yield, one attempts to maximise the associated ultimate load. According to the lower-bound theorem, the ultimate load is equal to or higher than the highest load found in this way and hence, the static method yields safe or lower-bound solutions for the actual ultimate load. Note that a state of stress obtained from a linear elastic analysis represents a statically admissible stress field since equilibrium and static boundary conditions are satisfied. Hence, although the elastically determined state of stress normally deviates from the actual state of stress in the structure, Chapter 3.1, design "based on the theory of elasticity" can be justified by the lower-bound theorem of limit analysis.

The kinematic or upper-bound method of the theory of plasticity is based on the upper-bound theorem. Starting from kinematically admissible states of deformation or failure mechanisms, one attempts to minimise the associated ultimate load. According to the upper-bound theorem, the ultimate load is equal to or lower than the lowest load found in this way and hence, the kinematic method yields unsafe or upper-bound solutions for the actual ultimate load.

In general, the establishment of a complete solution represents a difficult if not impossible task. However, the ultimate load can usually be bracketed to a satisfactory degree by establishing upper- and lower-bound solutions, Fig. 3.1 (c); see also Chapter 3.3.3.

3.2.3 Modified Coulomb Failure Criterion

In order to apply limit analysis methods to structural concrete, a suitable yield condition or failure criterion for the concrete must be assumed. Many different failure criteria have been proposed; relevant discussions are given by Chen [25] and Marti [82]. Overly sophisticated failure criteria for concrete are misleading since the scatter observed in tests on plain concrete is rather wide, Chapter 2.2, and modelling of concrete as a perfectly plastic material obeying the associated flow rule drastically idealises its actual behaviour anyway. Hence, the rather simple modified Coulomb failure criterion introduced by Chen and Drucker [22] will be adopted for the concrete in the present work. While this model depends only on three parameters, it is capable of adequately representing the behaviour of concrete for a wide range of stress combinations.

The Coulomb sliding failure criterion, $|\tau| + \sigma \tan\varphi - c = 0$, where τ and σ denote the shear and normal stresses acting on an arbitrary plane, has been applied to materials like soil and concrete for a long time. While this criterion adequately represents the behaviour of concrete for moderate compressive stresses, the concrete tensile strength is typically overestimated. Therefore, it was supplemented by a tension cut-off, resulting in the modified Coulomb failure criterion

$$|\tau| + \sigma \tan\varphi - c = 0 \qquad \text{and} \qquad \sigma - f_{ct} = 0 \qquad (3.4)$$

Fig. 3.2 (a) illustrates the modified Coulomb failure criterion in the stress plane. The sliding criterion can be expressed in terms of the principal stresses as

$$k\sigma_1 - \sigma_3 = f_c \qquad \text{where} \qquad k = \frac{1 + \sin\varphi}{1 - \sin\varphi} \quad \text{and} \quad f_c = \frac{2c\cos\varphi}{1 - \sin\varphi} \qquad (3.5)$$

Here f_c and f_{ct} are the uniaxial compressive and tensile concrete strengths, and c and φ denote the cohesion and the angle of internal friction of the concrete, respectively. In Eq. (3.5), σ_1 is the maximum and σ_3 the minimum principal stress; by exchanging the indices five additional equations corresponding to Eq. (3.5) are obtained. The resulting six equations define the Coulomb yield surface, an irregular hexagonal pyramid in the principal stress space. The pyramid is cut by three planes corresponding to Eq. $(3.4)_2$, and the failure envelope obtained in this way is the modified Coulomb yield surface, the projection and section of which with the (σ_1, σ_3)-plane are shown in Fig. 3.2 (b). The values of φ, f_c and f_{ct} are determined from tests which indicate that the angle of friction is nearly constant for all types of concrete at $\tan\varphi = 0.75$, corresponding to $k = 4$ and $c = f_c/4$.

The modified Coulomb failure criterion is a special case of a Mohr failure envelope, which in its most general form is of the type $|\tau| - g(\sigma) \leq 0$, Fig. 3.3 (c). As pointed out by Marti [82], the dissipation per unit volume of the concrete can be expressed as

$$dD = \boldsymbol{\sigma} \cdot \dot{\boldsymbol{\varepsilon}} = \tilde{c} \cot\tilde{\varphi} \dot{\varepsilon}_{(1)} \qquad (3.6)$$

for any isotropic material obeying a failure criterion of the Mohr type and following the theory of plastic potential; $\dot{\varepsilon}_{(1)} = \dot{\varepsilon}_1 + \dot{\varepsilon}_2 + \dot{\varepsilon}_3$ is the first invariant of the strain rate tensor

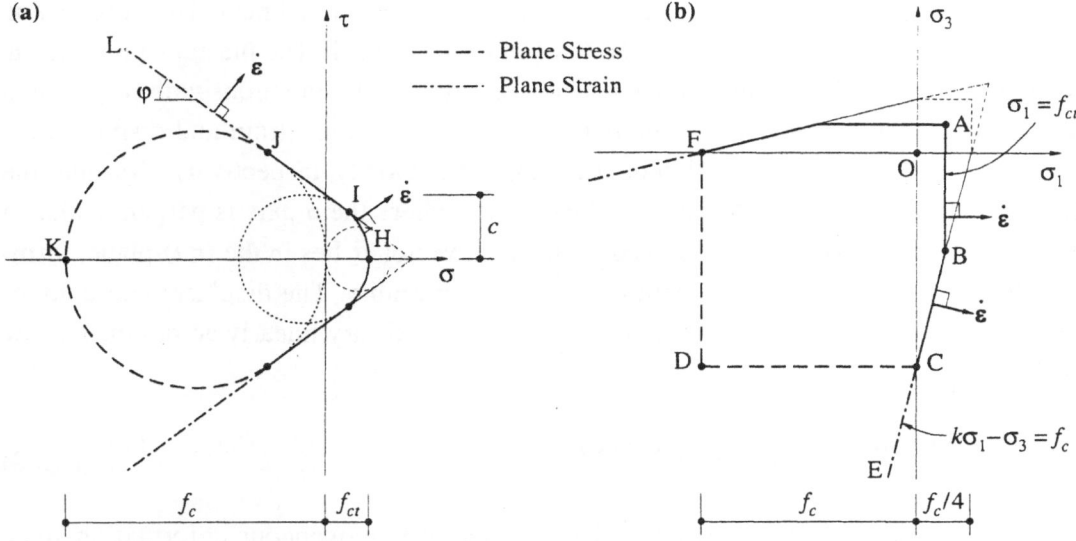

Fig. 3.2 – Modified Coulomb failure criterion for concrete: (a) stress plane; (b) principal stress space. Note: $\tan\varphi = 0.75$.

$\dot{\boldsymbol{\varepsilon}}$, and the values of $\tilde{c}$ and $\tilde{\varphi}$ correspond to the Coulomb yield surface circumscribing the actual failure criterion at the state of stress under consideration. For a Coulomb criterion, Eq. (3.6) was given by Chen [24]; it is directly found if the scalar multiplication of the stress and strain vectors is performed at the tip of the circumscribing Coulomb pyramid where $\boldsymbol{\sigma} = \tilde{c}\cot\tilde{\varphi}[1,1,1]$. For plane strain situations, $\dot{\varepsilon}_2 = 0$, Eq. (3.6) simplifies to

$$dD = \boldsymbol{\sigma}\cdot\dot{\boldsymbol{\varepsilon}} = \tilde{c}\cos\tilde{\varphi}\dot{\gamma}_{max} \tag{3.7}$$

where $\dot{\gamma}_{max} = \dot{\varepsilon}_1 - \dot{\varepsilon}_3$ is the maximum shear strain, see Eq. (3.5): $\sin\varphi = (k-1)/(k+1)$ and $\dot{\varepsilon}_1/\dot{\varepsilon}_3 = -k$.

3.2.4 Discontinuities

In most practical applications, discontinuous states of stress have to be considered, and failure mechanisms often consist of rigid parts separated by displacement discontinuities. The theorems of limit analysis remain valid for these cases if certain additional conditions are observed [43,136]. A detailed examination of possible states of stress and strain at discontinuities was given by Marti [82].

The lower-bound theorem of limit analysis remains valid for statically admissible states of stress containing discontinuities if the equilibrium conditions are fulfilled for elements on the stress discontinuities. Hence, only axial stresses acting parallel to the discontinuity may exhibit a jump across a stress discontinuity. In a state of plane stress, this means that the axial stress parallel to the discontinuity line may exhibit a jump, while the axial stress as well as the shear stress normal to the discontinuity have to be continuous.

Displacement discontinuities or failure surfaces are considered next. The notations $\dot{\varepsilon}$ and $\dot{u}$ will be used without superscript $^{(p)}$ as in Chapter 3.2.1. The plastic strain rates $\dot{\varepsilon}$ and the associated displacement rates $\dot{u}$ again represent differentiations with respect to a scalar rather than the physical time, and the terms dissipation, strain and displacement will be used rather than dissipation or strain rate and velocity, respectively. Consider the displacement discontinuity at Point P, Fig. 3.3 (a), where the n-axis is perpendicular to the plane of the discontinuity and the displacement vector $\dot{\delta}$ lies in the (n,t)-plane, forming the angle α with the t-axis parallel to the discontinuity. The displacement components $\dot{u}_n$ and $\dot{u}_t$, in the directions n and t, are assumed to vary linearly across the narrow failure zone of thickness d, i.e.

$$\dot{u}_n = \frac{|\dot{\delta}|n\sin\alpha}{d} \qquad \dot{u}_t = \frac{|\dot{\delta}|n\cos\alpha}{d} \tag{3.8}$$

This corresponds to constant strains within a zone of homogeneous deformation since $\dot{\varepsilon}_n = \partial\dot{u}_n/\partial n = (|\dot{\delta}|/d)\sin\alpha$, $\dot{\varepsilon}_t = \partial\dot{u}_t/\partial t = 0$ and $\dot{\gamma}_{tn} = \partial\dot{u}_n/\partial t + \partial\dot{u}_t/\partial n = (|\dot{\delta}|/d)\cos\alpha$ are independent of the coordinates n and t. Considering Mohr's circle of strains, Fig. 3.3 (b), one obtains

$$\dot{\varepsilon}_1 = \frac{|\dot{\delta}|}{2d}(1+\sin\alpha) \qquad \dot{\varepsilon}_3 = -\frac{|\dot{\delta}|}{2d}(1-\sin\alpha) \qquad \theta_3 = \theta_t + \frac{\alpha}{2} - \frac{\pi}{4} \tag{3.9}$$

At the limit of d tending to zero, both $\dot{\varepsilon}_1$ and $\dot{\varepsilon}_3$ tend to infinity while $\dot{\varepsilon}_2$ has a finite value, indicating a state of plane strain at the discontinuity since $\dot{\varepsilon}_2/\dot{\varepsilon}_1 \to 0$ as well as $\dot{\varepsilon}_2/\dot{\varepsilon}_3 \to 0$. This is particularly useful if failure is constrained to a certain plane in structures capable of resisting stresses in the $y \equiv 2$-direction, e.g. in a construction joint. If the failure zone has a finite thickness $d > 0$, a state of plane stress may have to be considered. In either case, $y \equiv 2$ is a principal strain direction and consideration of a failure line in the (n,t)-plane is sufficient. The principal directions 1 and 3 bisect the angles between the parallel to the discontinuity, (I), and the normal to the displacement direction, (II). In the so-called slip-line or characteristic directions (I,II) pure shear strains occur, and the principal strains $\dot{\varepsilon}_1$ and $\dot{\varepsilon}_3$ have opposite signs except for the special case of $\alpha = \pi/2$.

The upper-bound theorem remains valid for discontinuous states of strain if the dissipation in the discontinuities is taken into account when calculating the internal energy dissipation. As shown by Marti [82], any yield condition of an isotropic material can be substituted by its associated Mohr envelope without loss of generality when considering failure surfaces, i.e., displacement discontinuities in plane strain, since the principal strains $\dot{\varepsilon}_1$ and $\dot{\varepsilon}_3$ have opposite signs. Setting $\dot{\gamma}_{max} = \dot{\varepsilon}_1 - \dot{\varepsilon}_3 = |\dot{\delta}|/d$, Fig. 3.3 (b), the dissipation per unit area of a failure surface follows from Eq. (3.7) as

$$dD \cdot d = \boldsymbol{\sigma} \cdot \dot{\boldsymbol{\varepsilon}} = |\dot{\delta}|\tilde{c}\cos\tilde{\varphi} \tag{3.10}$$

where the values of $\tilde{c}$ and $\tilde{\varphi}$ again correspond to the Coulomb yield surface circumscribing the actual failure criterion at the state of stress under consideration, i.e., $\tilde{\varphi} = \alpha$ as illustrated in Fig. 3.3 (c).

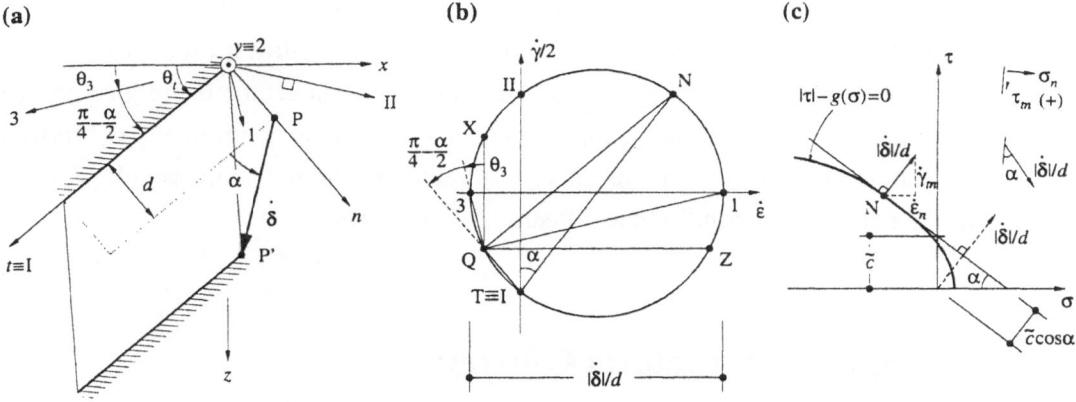

Fig. 3.3 – Displacement discontinuity: (a) notation; (b) Mohr's circle of strains; (c) arbitrary Mohr envelope in stress plane.

Expressions for the dissipation in displacement discontinuities can also be established by explicitly calculating the scalar product of the stress and the strain vectors and by using the fact that the stress vector is determined by the assumed yield criterion for the concrete and the associated flow rule. Chen and Drucker [22] showed that the dissipation per unit area of a failure surface in a material obeying a modified Coulomb failure criterion is equal to

$$dD \cdot d = |\dot{\boldsymbol{\delta}}| \left[\frac{f_c}{2}(1 - \sin\alpha) + f_{ct}\frac{\sin\alpha - \sin\varphi}{1 - \sin\varphi} \right] \qquad \text{for} \quad \varphi \leq \alpha \leq \frac{\pi}{2} \qquad (3.11)$$

In the case of plane strain no state of stress corresponds to a deformation with $\alpha < \varphi$; this situation could only occur if a non-associated flow rule were adopted. For $\alpha = \varphi$ and $\alpha = \pi/2$, the dissipation is independent of the tensile and compressive strength, respectively. If the tensile strength is neglected, one gets

$$dD \cdot d = |\dot{\boldsymbol{\delta}}| \frac{f_c}{2}(1 - \sin\alpha) \qquad \text{for} \quad \varphi \leq \alpha \leq \frac{\pi}{2} \qquad (3.12)$$

For $\alpha = \pi/2$ (pure opening) no dissipation occurs in the concrete; Müller [109] introduced the term "collapse crack" for this case.

The above expressions can also be applied to discontinuous states of deformation in walls subjected to a state of plane stress. In this case, the condition $\alpha \geq \varphi$ does not apply; Line JK and Point K in Fig. 3.2 (a), i.e., Point C and Line CD Fig. 3.2 (b), respectively, correspond to situations where $\alpha < \varphi$. For $\varphi \leq \alpha \leq \pi/2$, the dissipation is determined by Eq. (3.11) and for $-\pi/2 \leq \alpha \leq \varphi$, it follows from Eq. (3.12). If the tensile strength of the concrete is neglected, the dissipation is determined by Eq. (3.12) for any value of α, i.e., for $-\pi/2 \leq \alpha \leq \pi/2$.

In order to apply the uniqueness theorem to discontinuous states of stress and strain, the compatibility condition has to be generalised [82,136] as follows: on every element of a discontinuity, the vector of stresses $\boldsymbol{\sigma}$ and the displacement vector $\dot{\boldsymbol{\delta}}$ are located in a plane orthogonal to the discontinuity, Fig. 3.3 (a), and the stress components σ_n and τ_{tn} follow from the inclination α of the displacement vector $\dot{\boldsymbol{\delta}}$ and the Mohr envelope associated with the actual yield condition as illustrated in Fig. 3.3 (c).

3.3 Application to Structural Concrete

3.3.1 Reinforcement and Bond

In the application of limit analysis methods to structural concrete, the reinforcing bars are typically assumed to be perfectly plastic, resisting only axial stresses. Furthermore, one assumes that the reinforcing bars are well-distributed, such that their effect can be modelled by equivalent stresses. In addition, since the bond between reinforcement and concrete is assumed to be perfect, no relative displacements between steel and concrete can occur, and any change of stresses in the reinforcement below the yield stress can be realised with an infinitely small bond or development length. In summary, the discrete reinforcing bars are substituted by infinitely thin, perfectly bonded fibres, and cracks with a vanishing spacing are assumed. Prestressed reinforcement is treated in essentially the same manner.

Apart from the assumption of perfectly plastic reinforcement, these idealisations are quite crude. In a real structure, reinforcing bars are not infinitely thin, and considerable transverse shear stresses may occur in the reinforcement ("dowel action"). Bond stresses are limited by the bond strength, resulting in finite development lengths. The crack spacings are not infinitely small and tension stiffening effects occur. On the other hand, the analysis of a structure is simplified to a great extent by these assumptions, and their influence on the ultimate load is often negligible.

In the application of limit analysis methods, concrete and reinforcement are typically considered together as a structural concrete continuum. The resistance of structural concrete is given by the linear combination of the resistances of concrete and reinforcement. If the yield surfaces of concrete and reinforcement are known in terms of generalised stresses, the yield surface of structural concrete can be interpreted geometrically as the surface obtained when translating the yield surface of the plain concrete such that its origin moves on the yield locus of the reinforcement, or vice versa [122,80,109]. Alternatively, concrete and reinforcement can be considered separately; in this case, the stresses transferred from steel to concrete are taken into account as distributed external forces in the analysis of the concrete continuum.

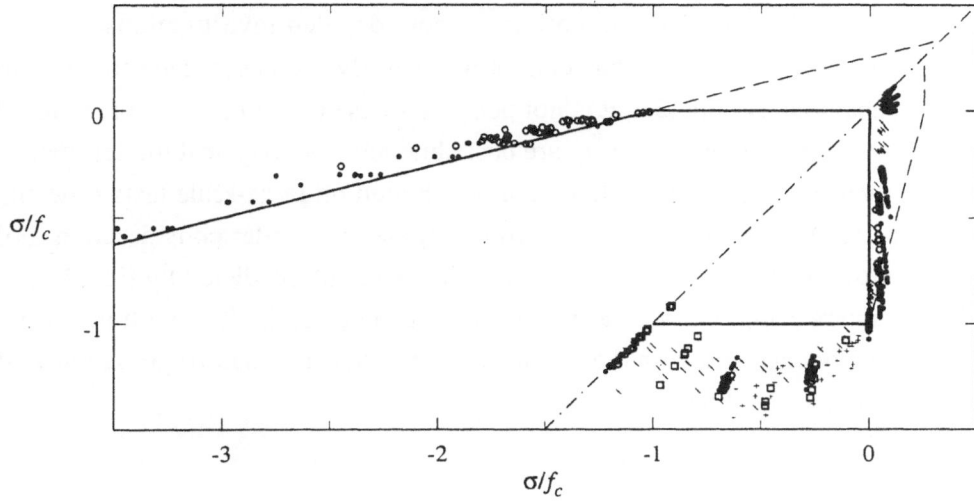

Fig. 3.4 – Modified Coulomb failure criterion: comparison with tests in plane stress (lower right) and plane strain (upper left). For legends see Figs. 2.6 and 2.7, respectively.

3.3.2 Effective Concrete Compressive Strength

Fig. 3.4 compares the test results presented in Chapter 2.2 with the modified Coulomb failure criterion for $\tan\varphi = 0.75$ and $f_{ct} = 0$. For plane strain, i.e., triaxial compression with $\sigma_1 = \sigma_2$, the modified Coulomb yield criterion is equivalent to Eq. (2.5), while for plane stress and $f_{ct} = 0$, it corresponds to the square failure criterion, Chapter 2.2.3. Although the test results agree well with the modified Coulomb failure criterion, the idealisation of concrete as a perfectly plastic material following the theory of plastic potential and obeying a modified Coulomb failure criterion is debatable in the light of its brittleness and strain-softening behaviour. However, if the limited ductility of concrete is taken into account by choosing an appropriately reduced value for the effective concrete compressive strength f_c, the perfectly plastic idealisation may still be adopted as a basis for the calculation of collapse loads of concrete structures.

If upper-bound solutions for the ultimate load according to limit analysis are established, the value of the effective concrete compressive strength f_c should be chosen such that the total energy dissipated in the concrete and the reinforcement at collapse is accurately represented. In the design stage, concrete dimensions are often determined from rather conservative values of f_c in order to avoid collapses governed by concrete crushing. In the resulting "under-reinforced" structures, most of the energy dissipation takes place in the reinforcement and hence, the exact assessment of the dissipation in the concrete is unnecessary. Eq. (2.26) represents a suitable estimate of the concrete compressive strength in such situations.

In the evaluation of existing structures and in the design of weight-sensitive structures such as long-span bridges or offshore platforms, more detailed investigations are often required since the concrete dimensions cannot be liberally increased. Due to the complexity of the behaviour of concrete, it is not possible to use the same value of f_c for all possible situations. Realistic values of f_c are best obtained from physical models that account for the governing parameters and that are calibrated on large-scale tests reflecting the actual states of stress and strain in the structural element under consideration. Following these lines, the effective concrete compressive strength predicted by Eq. (2.27) is validated in Chapters 5 and 6 as well as in Appendix B by a comparison with the available experimental evidence and with the results of comprehensive load-deformation analyses obtained from a refined model.

3.3.3 Stress Fields and Failure Mechanisms

The theorems of limit analysis can be directly applied to the analysis and design of concrete structures if the assumptions outlined in Chapter 3.3.1 are made and a suitable value of the effective concrete compressive strength f_c is chosen. A comprehensive account of stress field and failure mechanism approaches for plane stress situations is given in [143].

Stress field solutions result in safe designs and an appropriate detailing since the flow of forces is followed consistently throughout the structure. Usually, some resistances such as tensile stresses in the concrete and membrane action in slabs are neglected. Furthermore, the same residual state of stress is typically superimposed on the elastically determined state of stress for all load-cases in statically indeterminate structures, while higher ultimate loads would generally result from the superposition of different residual states of stress on each load-case [87,89,143]. A certain conservatism is thus inherent to stress-field solutions.

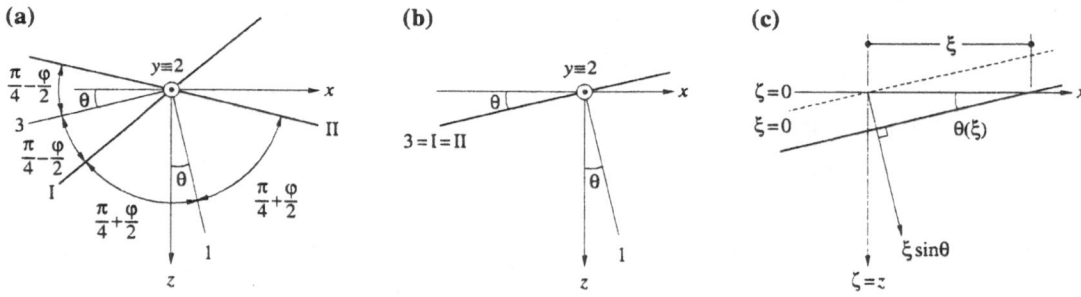

Fig. 3.5 – Characteristic directions: (a) general hyperbolic case; (b) parabolic special case $\varphi = \pi/2$; (c) parametric representation of straight characteristics for $\varphi = \pi/2$, no distributed forces acting on the concrete.

When analysing existing structures or checking reinforcement layouts determined by computer programs, the consideration of failure mechanisms offers the advantage of quickly verifying essential dimensions and details even in complicated cases. Actual collapse mechanisms often exhibit discontinuous displacement fields; displacement discontinuities, Chapter 3.2.4, allow to treat such cases and to establish upper-bounds for the collapse load. The resulting upper-bound values of the bearing capacity may also be used to estimate the degree of conservatism of a stress field design. Often, certain resistances are neglected and some kinematic conditions are violated in establishing failure mechanisms. Ultimate loads observed from tests may thus be higher than the lowest theoretical upper-bound solution.

3.3.4 Characteristic Directions in Plane Stress

For plane stress situations, the ultimate state of stress in a perfectly plastic material following the associated flow rule satisfies a system of quasi-linear partial differential equations [135,109,82]. Marti [82] analysed the governing equations for a modified Coulomb failure criterion. The necessary calculations are given in Appendix A for the coordinate system and sign conventions used in the present thesis; some results are outlined below. Knowing the properties of the characteristic directions in a displacement discontinuity, Chapter 3.2.4, the same results could be obtained directly from the compatibility condition, Chapters 3.2.2 and 3.2.4, rather than from the calculations outlined in Appendix A.

In general, the system of differential equations is hyperbolic, and the directions of the characteristics are determined by

$$\frac{dz}{dx} = \cot\left[\theta \pm \left(\frac{\varphi}{2} + \frac{\pi}{4}\right)\right] \tag{3.13}$$

see Fig. 3.5 (a). In Eq. (3.13), θ is the inclination of the principal compressive stress direction with respect to the x-axis, and φ corresponds to the Coulomb yield surface circumscribing the failure criterion at the state of stress under consideration, Chapter 3.2.3. The characteristics determined by Eq. (3.13) have several special properties [135]; they are inextensible and constitute potential lines of discontinuity in the state of strain, i.e., slip lines according to Chapter 3.2.4. In complete solutions, the displacement vector is perpendicular to the characteristic direction which is not the failure line.

For the special case $\varphi = \pi/2$, i.e., Point H in Fig. 3.2 (a) and Line AB in Fig. 3.2 (b), the system of differential equations is parabolic, and both directions of the characteristics coincide with the principal compressive stress direction,

$$\frac{dz}{dx} = -\tan\theta \tag{3.14}$$

In this case which is illustrated in Fig. 3.5 (b) the directions of the characteristics are potential collapse cracks, Chapter 3.2.4, and such displacement discontinuities may coincide with discontinuities in the state of stress. This is not possible in complete solutions for the general, hyperbolic case.

If no distributed forces act on the concrete, the directions of the characteristics determined by Eq. (3.14) constitute a one-parameter set of straight lines which can be represented by an expression of the type

$$\cot\theta = g(x + z\cot\theta) \tag{3.15}$$

where g denotes an arbitrary function to be determined from the boundary conditions for $\cot\theta$ along $z = 0$, see Fig. 3.5 (c). Introducing skew coordinates $\xi = x + z\cot\theta$, $\zeta = z$, one obtains

$$\cot\theta = g(\xi) \tag{3.16}$$

The directions of the characteristics are thus determined by their inclination along $\zeta = z = 0$, Fig. 3.5 (c). The envelope of all characteristics can be found by differentiation of $\xi = x + z\cot\theta$ with respect to ξ and elimination of the parameter ξ.

The difference between the principal tensile and compressive stresses along a straight characteristic determined by Eq. (3.16) is inversely proportional to the distance from the point where the characteristic is tangent to the envelope, i.e., the difference between the principal stresses varies hyperbolically along the characteristic

$$\sigma_1 - \sigma_3 = \frac{(\sigma_1 - \sigma_3)\big|_{\zeta=0}}{1 - \zeta\dfrac{d}{d\xi}[\cot\theta(\xi)]} \tag{3.17}$$

The straight characteristics determined by Eq. (3.16) remain straight and unstrained, i.e., if no distributed forces act on the concrete and $\varphi = \pi/2$, the deformations at the ultimate state are composed of translations of the characteristics and displacements perpendicular to the characteristics, varying linearly along them [82].

4   Previous Work on Plane Stress Problems

4.1   General

Rather than attempting to provide a complete review of the many previous investigations on plane stress in structural concrete, the present chapter concentrates on physical models which significantly influenced today's knowledge in the field of membrane shear behaviour and it describes the relationships between the different approaches. After a brief historical account some particularly important models are examined in detail. While only little reference is made to investigations on torsion in structural concrete, it should be kept in mind that torsion research has substantially contributed to the present state-of-the-art of membrane shear behaviour [90].

Truss or strut and tie models have been used for following the flow of internal forces in reinforced concrete structures for more than a century. Hennebique's construction method [41], first patented in 1892 and supplemented by a patent for vertical stirrups in 1893, already introduced a truss analogy [129] to calculate the stirrup forces. Ritter [129] gave a concise explanation of the truss model, but he argued that the stirrups, rather than acting as tensile posts of a truss, might merely counteract the principal tensile stresses in the concrete and serve as a measure against premature cracking. In order to clarify the subject, he recommended to perform pertinent tests. During the first two decades of this century, Mörsch [104,106] advanced the classical 45-degree truss model and initiated several test series on beams subjected to flexure and shear to verify the model. Through careful test observations, Mörsch knew that diagonal cracks in the web may become considerably flatter with decreasing amounts of web reinforcement, resulting in smaller stirrup forces than according to the 45-degree truss model. However, he thought that it was practically impossible to determine the actual inclination of the diagonal cracks and recommended to continue using the simple and conservative 45-degree truss model. Mörsch's 45-degree truss model was the starting point of most theoretical and experimental investigations on shear in structural concrete, Fig. 4.1, and it has been adopted by many codes of practice as the basis of their shear and torsion design provisions.

Extensive test programs [3] on beams subjected to flexure, shear and normal forces confirmed that the 45-degree truss model is overly conservative for small amounts of web reinforcement. Variable angle truss models, i.e., truss models with flatter inclinations of the compression diagonals, can overcome this discrepancy. Such approaches were used well before applying the theory of plasticity to structural concrete [72,76]. They are presented as limit analysis methods in Fig. 4.1 because they were put on a sound physical basis through the establishment of the lower-bound theorem of limit anal-

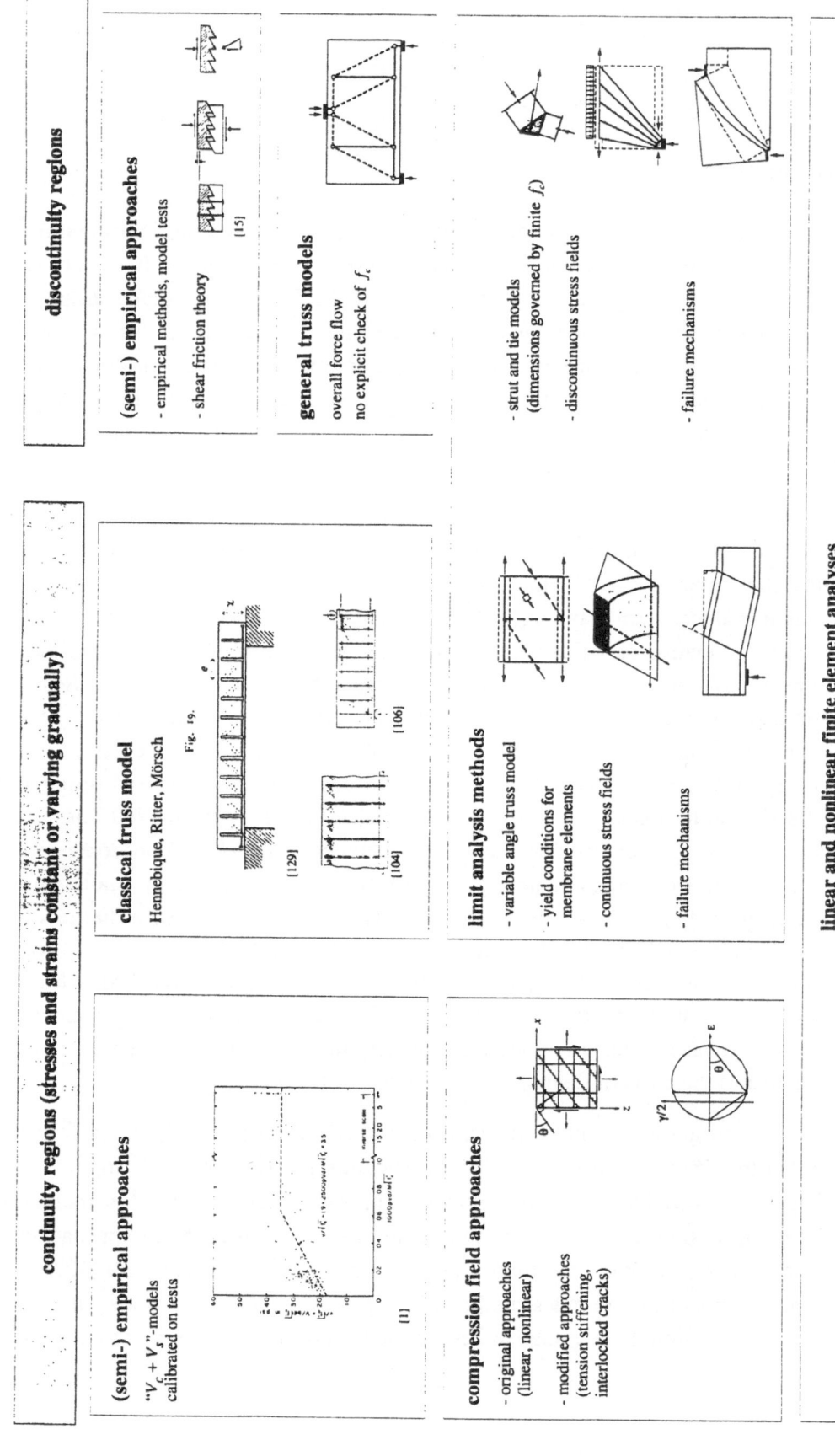

Fig. 4.1 – Shear design methods for structural concrete.

ysis. In Fig. 4.1, truss models are separated into general truss models whose primary concern is the overall force flow [138], and strut and tie models which consider finite values of the concrete compressive strength and hence finite dimensions of the compression struts and nodal zones [82,113,107,89]. Strut and tie models represent a special case of discontinuous stress fields within limit analysis.

Rather than assuming truss models with flatter inclinations of the compression diagonals, the shear resistance exceeding that predicted by the 45-degree truss model has been attributed to the flexural compression zone of the beams and to bending action of so-called concrete cantilevers bounded by the diagonal shear cracks [62,63]. In these semi-empirical approaches, a so-called concrete contribution V_c to the shear resistance is added to the so-called steel contribution V_s, where V_s = shear resistance according to the 45-degree truss model. However, as pointed out by Fenwick and Paulay [50], the assumption that V_c is carried in the manner outlined above is incorrect for girders with web reinforcement where stresses transferred across the diagonal shear cracks by aggregate interlock result in flatter inclinations of the principal compressive stress direction. Hence, if the tensile strength of concrete is neglected, V_c merely corresponds to the difference between the shear resistance according to the 45-degree truss model and that determined from a truss model with flatter inclination of the compression diagonals.

"$V_c + V_s$"-approaches are still commonly used in North America [1,2,3,4]. Unlike truss models neglecting the tensile strength of concrete, they can predict the shear strength of members without web reinforcement. While the semi-empirical "$V_c + V_s$"-approaches may simplify the treatment of well-defined standard problems, their applicability is restricted to the range covered by the underlying test results. Hence, "$V_c + V_s$"-approaches will not be further commented here.

4.2 Limit Analysis Methods

The application of the theory of plasticity to structural concrete subjected to in-plane shear and normal stresses has led to a unification of the design methods for beams, walls, slabs and shells [95]. This chapter reviews limit analysis approaches for membrane elements and beams; more extensive information can be found in an IABSE state-of-the-art report [60] and in monographs by Thürlimann et al. [152] and Nielsen [113].

A considerable number of failure mechanisms or upper-bound solutions as well as complete solutions have been presented [109,82,113]. Typically, failure mechanisms involving discrete failure lines have been studied; the dissipation in the concrete follows from Eqs. (3.11) and (3.12), and the assessment of the dissipation in the reinforcement is straightforward. Upper-bound solutions are particularly useful for the analysis of existing structures and for checking essential dimensions and details of a design. For an overview of failure mechanisms in plane stress situations, see [143].

4.2.1 Stress Fields

Stress field approaches apply to any structural geometry and loading. They serve as simple and transparent design tools that permit visualising the force flow throughout a structure, allowing for a consistent dimensioning and detailing. Generally, stress fields should be applied in conjunction with a well distributed minimum reinforcement in the regions where no main reinforcement is required. Strut and tie models consisting of individual tension and compression members are suitable for most design situations. Starting from simple strut and tie models, more sophisticated stress fields that closely represent the actual behaviour of a structure can often be developed [82,85,107,89,143]. Sigrist [144] successfully applied such discontinuous stress fields to problems of the deformation capacity of girders. Discontinuous stress fields are particularly useful in the design of girders with flanged cross-sections [84,86,87,92]; as illustrated below, they indicate the necessary amount, the correct position and the required dimensioning and detailing of the longitudinal as well as the transverse reinforcement. No special shift rules for the longitudinal reinforcement are necessary, concrete dimensions and support widths can be checked, and development and anchorage lengths can easily be accounted for.

Fig. 4.2 illustrates the application of discontinuous stress fields to the design of a single-span girder with an overhang. For comparison purposes, the truss model representing the resultant forces of the stress field is also shown. The discontinuous stress field illustrated in Fig. 4.2 (b) consists of parallel stress bands (parallelogram-shaped regions with constant inclination θ of the uniaxial compressive stress field in the web concrete) and fans (triangular regions with variable θ, at locations where concentrated loads are introduced). The fans shown in Fig. 4.2 (b) are called centred since all trajectories meet in one point (which might also be located away from the chords). Starting from the points of zero shear force, chord and stirrup forces, Fig. 4.2 (c), can be determined. The stirrup forces f_w (per unit length) are easily obtained from considering sections along the inclined boundaries of parallel bands or fans where $f_w = \text{constant}$. For loads acting on the top chord, the product $f_w d_v \cot\theta$ lies within the shear force diagram; this has been called staggering effect [86,87]. Loads q_{inf} acting on the bottom chord would have to be suspended by additional stirrup forces of equal amount, $\Delta f_w = q_{inf}$. The forces in the top and bottom chords are related to the applied loads and the stirrup forces by

$$\frac{d}{dx}F_{sup} = -(q + f_w)\cot\theta \qquad\qquad \frac{d}{dx}F_{inf} = f_w\cot\theta \qquad\qquad (4.1)$$

see Fig. 4.2 (d). According to Eq. (4.1), the chord forces vary linearly along parallel stress bands (constant value of $\cot\theta$) and parabolically along the boundaries of centred fans (constant value of the effective depth d_v and linear variation of $\cot\theta$) for constant loads and stirrup forces. The concrete compressive stresses can be determined from

$$-\sigma_{c3} = (q + f_w)(1 + \cot^2\theta)/b_w \qquad\qquad -\sigma_{c3} = f_w(1 + \cot^2\theta)/b_w \qquad\qquad (4.2)$$

along the top and bottom chord, respectively, Fig. 4.2 (d). These expressions allow checking the web thickness b_w. Concrete stresses are constant in the parallel stress bands

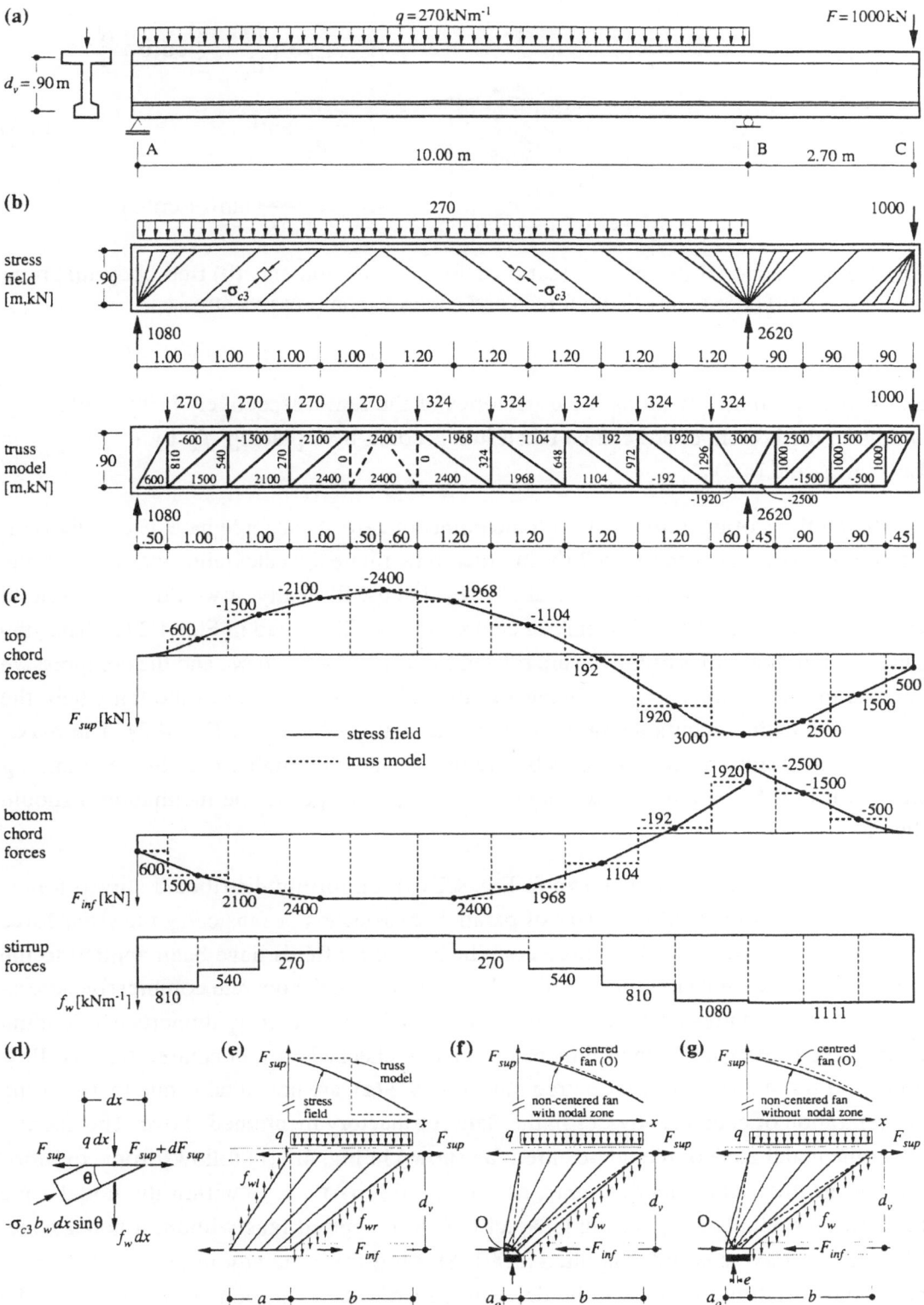

Fig. 4.2 – Stress field analysis: (a) geometry and loading; (b) stress field and truss model of resultant forces; (c) resulting chord and stirrup forces; (d) forces acting on top chord; (e) centred fan; (f) non-centred fan with nodal zone; (g) non-centred fan without nodal zone.

(a) **(b)** **(c)**

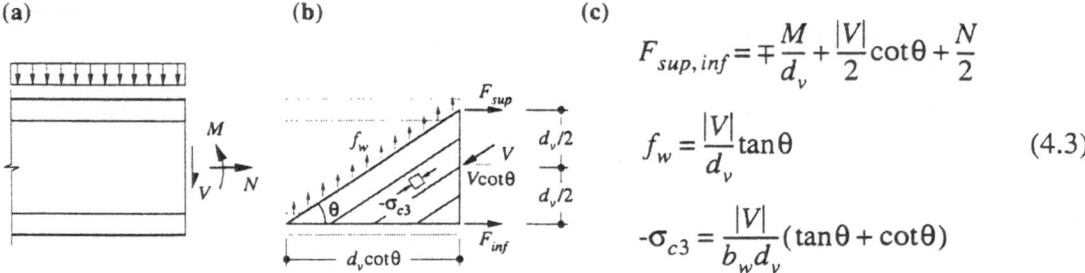

$$F_{sup,inf} = \mp \frac{M}{d_v} + \frac{|V|}{2}\cot\theta + \frac{N}{2}$$

$$f_w = \frac{|V|}{d_v}\tan\theta \qquad\qquad (4.3)$$

$$-\sigma_{c3} = \frac{|V|}{b_w d_v}(\tan\theta + \cot\theta)$$

Fig. 4.3 – Sectional design: (a) sectional forces and moment; (b) free-body diagram; (c) chord and stirrup forces and concrete compressive stresses.

and vary hyperbolically along the trajectories of the fans, respectively [109]. Loads q_{inf} acting on the bottom chord can be taken into account by replacing f_w by $(f_w - q_{inf})$ in Eqs. (4.1)$_2$ and (4.2)$_2$.

The portions of the girder between the points of zero shear force have been subdivided into equal distances in Fig. 4.2 (b) in order to facilitate the calculations; as a result, the stress field primarily consists of parallel stress bands. Basically, it would be possible to choose a stress field consisting only of centred fans as illustrated in Fig. 4.2 (e). The stirrup forces on both sides of the fan are related by $f_{wl} = (q + f_{wr})b/a$, and the compressive stresses in the concrete $-\sigma_{c3}$ are highest in the bottom right corner of the fan where the flattest inclination occurs along with the higher value of f_w, see Eq. (4.2). The maximum compressive stresses in the web may be considerably higher than in the adjoining trajectories of the parallel stress band [109]; abrupt changes in the inclination θ should thus be avoided.

In the simple stress field shown in Fig. 4.2 (b), uniformly distributed stirrup forces acting along any inclined boundary of parallel stress bands or fans carry the shear force at the corresponding location; more complicated stress fields have been applied to the evaluation of existing structures [146]. Also, infinitely high concrete compressive stresses result at the centres of the fans shown in Fig. 4.2 (b). In reality, supports and loading plates have finite widths; their dimensions can be checked by non-centred fans as illustrated in Figs. 4.2 (f) and (g). Centred fans, Fig. 4.2 (e), are not suitable due to the strong concentration of concrete stresses in the flattest trajectory mentioned above. The geometry of the nodal zone of the non-centred fan shown in Fig. 4.2 (f) follows from equilibrium and geometrical considerations, assuming $-\sigma_{c3} = -\sigma_{c1} = f_c$ within the nodal zone [82]; further details are given in Chapter 6.3. Non-centred fans without a nodal zone, Fig. 4.2 (g), have also been applied [109]. Starting from the adjoining parallel stress band, the variation of $\cot\theta$ can be determined from the condition $-\sigma_{c3} = f_c$ along the support. This type of fan is of limited practical value since for equal values of f_c, it requires longer supports than non-centred fans with a nodal zone, Fig. 4.2 (f).

The chord forces resulting from non-centred fans are lower than those determined from a fan centred in point O, Figs. 4.2 (f) and (g) (as long as the height of the nodal zone in Fig. 4.2 (f) fits inside the bottom chord; otherwise, an iterative calculation using a reduced value of d_v is necessary). Thus, the chord forces can be safely determined from the simpler centred fan. Non-centred fans are only applied to check the dimensions of supports and loading plates. The effective concrete compressive strength f_c to be used at supports is not easy to assess; further details are given in Chapter 6.3.

If all static and geometric quantities vary only gradually along the girder axis, a "sectional" design approach can be developed, Fig. 4.3. Note that the sectional forces M and N have been reduced to mid-depth of the web which only occasionally coincides with the girder axis. Strictly speaking, the stirrups resulting from Eq. (4.3) must be provided along a distance $d_v \cot\theta$, and the approach is thus not purely sectional. The inclination θ can be chosen in design and hence, the relations illustrated in Fig. 4.3 have been called variable angle truss model. Certain limitations, such as $0.5 \le \cot\theta \le 2$, are usually imposed on θ in order to avoid premature failures due to concrete crushing or steel rupture.

4.2.2 Yield Conditions for Membrane Elements

Rather than determining the reinforcement from a discontinuous stress field solution, admissible stress fields are often obtained from finite element analyses in design practice. Yield conditions for membrane elements [112,109,82] allow dimensioning of the reinforcement in such cases. Consider an orthogonally reinforced panel element subjected to homogeneous in-plane stresses σ_x, σ_z and τ_{xz}, Fig. 4.4 (a). Reinforcement ratios and yield strengths in tension and compression are denoted by ρ, f_y and f_y', respectively. Thus, in the space of the stress components, the rectangular yield locus of the reinforcement shown in Fig. 4.4 (b) is obtained. The combined no-tension and no-crushing criterion $0 \ge \sigma \ge -f_c$ for plain concrete corresponds to two conical surfaces described by the equations $\tau_{xz}^2 = \sigma_x \sigma_z$ and $\tau_{xz}^2 = (f_c + \sigma_x)(f_c + \sigma_z)$, Fig. 4.4 (c), where f_c = effective concrete compressive strength, see Chapter 3.3.2.

The yield surface of the reinforced concrete panel is obtained by translating the yield surface of the plain concrete such that its origin moves within the yield locus of the reinforcement, or vice versa, see Chapter 3.3. The envelope of all such stress combinations is the yield surface shown in Fig. 4.4 (d). It is described by the yield criteria

$$
\begin{aligned}
\Phi_1 &= \tau_{xz}^2 - (\rho_x f_{yx} - \sigma_x)(\rho_z f_{yz} - \sigma_z) &&= 0 \\
\Phi_2 &= \tau_{xz}^2 - (f_c - \rho_z f_{yz} + \sigma_z)(\rho_z f_{yz} - \sigma_z) &&= 0 \\
\Phi_3 &= \tau_{xz}^2 - (\rho_x f_{yx} - \sigma_x)(f_c - \rho_x f_{yx} + \sigma_x) &&= 0 \\
\Phi_4 &= \tau_{xz}^2 - f_c^2/4 &&= 0 \qquad (4.4) \\
\Phi_5 &= \tau_{xz}^2 + (\rho_x f_{yx}' + \sigma_x)(f_c + \rho_x f_{yx}' + \sigma_x) &&= 0 \\
\Phi_6 &= \tau_{xz}^2 + (f_c + \rho_z f_{yz}' + \sigma_z)(\rho_z f_{yz}' + \sigma_z) &&= 0 \\
\Phi_7 &= \tau_{xz}^2 - (f_c + \rho_x f_{yx}' + \sigma_x)(f_c + \rho_z f_{yz}' + \sigma_z) &&= 0
\end{aligned}
$$

characterising the seven regimes identified in Fig. 4.4 (e). Points on the yield surface correspond to potential failure stress combinations. According to the flow rule

$$\dot{\varepsilon}_x = \kappa\frac{\partial\Phi}{\partial\sigma_x} \qquad \dot{\varepsilon}_z = \kappa\frac{\partial\Phi}{\partial\sigma_z} \qquad \dot{\gamma}_{xz} = \kappa\frac{\partial\Phi}{\partial\tau_{xz}} \tag{4.5}$$

the plastic strain rates $\dot{\varepsilon}_x$, $\dot{\varepsilon}_z$, and $\dot{\gamma}_{xz}$ are proportional to the components of the outward normal to the yield surface in the stress point under consideration, Chapter 3.2.1; the symbol κ denotes an arbitrary non-negative factor. For stress points satisfying more than one of the criteria (4.4) the flow rule (4.5) has to be applied separately for each regime and the associated strain rates must be superimposed. Knowing the strain rates, associated Mohr's circles can be drawn as shown in Fig. 4.4 (f) and the inclination θ of the principal compressive direction 3 to the x-axis can be determined from

$$\cot 2\theta = \frac{\dot{\varepsilon}_z - \dot{\varepsilon}_x}{\dot{\gamma}_{xz}} \quad \text{or} \quad \cot\theta = \frac{\dot{\varepsilon}_z - \dot{\varepsilon}_x}{\dot{\gamma}_{xz}} + \sqrt{\left(\frac{\dot{\varepsilon}_z - \dot{\varepsilon}_x}{\dot{\gamma}_{xz}}\right)^2 + 1} \tag{4.6}$$

which yields the following expressions for the seven regimes identified in Fig. 4.4 (e):

$$\Phi_1 : \quad \cot^2\theta = (\rho_x f_{yx} - \sigma_x)/(\rho_z f_{yz} - \sigma_z)$$

$$\Phi_2 : \quad \cot^2\theta = (f_c - \rho_z f_{yz} + \sigma_z)/(\rho_z f_{yz} - \sigma_z)$$

$$\Phi_3 : \quad \cot^2\theta = (\rho_x f_{yx} - \sigma_x)/(f_c - \rho_x f_{yx} + \sigma_x)$$

$$\Phi_4 : \quad \cot^2\theta = 1 \tag{4.7}$$

$$\Phi_5 : \quad \cot^2\theta = -(\rho_z f'_{yx} + \sigma_x)/(f_c + \rho_x f'_{yx} + \sigma_x)$$

$$\Phi_6 : \quad \cot^2\theta = -(f_c + \rho_z f'_{yz} + \sigma_z)/(\rho_z f'_{yz} + \sigma_z)$$

$$\Phi_7 : \quad \cot^2\theta = (f_c + \rho_x f'_{yx} + \sigma_x)/(f_c + \rho_z f'_{yz} + \sigma_z)$$

Regime 1 corresponds to under-reinforced structures characterised by yielding of both reinforcements at failure. This regime is particularly important in design practice. Rearranging Eq. (4.4)$_1$ one obtains the requirements

$$\rho_x f_{yx} \geq \sigma_x + k|\tau_{xz}| \qquad \rho_z f_{yz} \geq \sigma_z + k^{-1}|\tau_{xz}| \tag{4.8}$$

where $k = \cot\theta$. As indicated in Figs. 4.4 (d) and (g), Eq. (4.8) provides a parametric representation of lines of equal shear stress τ_{xz}. Most design codes impose similar limitations on $k = \cot\theta$ as on the inclination of the compressive stresses in the variable angle truss model; typically, $0.5 \leq k \leq 2$ is allowed. These limitations are also indicated in Fig. 4.4 (g) along with the yield regimes discussed above. In design practice, $k = 1$ is often chosen; while this assumption results in safe designs, all other feasible designs are disregarded. Eq. (4.8) is only valid in Regime 1, i.e., if $f_c \geq \rho_x f_{yx} + \rho_z f_{yz} - (\sigma_x + \sigma_z)$. Otherwise, the concrete crushes and one (Regimes 2 and 3) or both reinforcements (Regime 4) remain elastic. In particular, if the reinforcement in the x-direction is stronger than that in the z-direction, i.e., $(\rho_x f_{yx} - \sigma_x) > (\rho_z f_{yz} - \sigma_z)$, failure is governed by crushing of the concrete and yielding of the z-reinforcement, Regime 2. This type of collapse is known as web crushing failure [16]. Fig. 4.4 (h) illustrates the corresponding shear resistances along with the limitations $0.5 \le k \le 2$.

The yield conditions (4.4) were first given by Nielsen [112] for isotropic reinforcement. Müller [109] presented a comprehensive discussion of the static and kinematic conditions for all yield regimes as well as yield criteria for generally reinforced membrane elements. The yield locus of a skew reinforcement mesh is indicated in Fig. 4.4 (i); the action of a reinforcement layer inclined at an angle β with respect to the x-axis can be represented by a vector with the components $\rho f_y \cos^2\beta$, $\rho f_y \sin^2\beta$ and $-\rho f_y \sin\beta\cos\beta$ in the directions of σ_x, σ_z and τ_{xz}, respectively. The parallelogram-shaped yield locus shown in Fig. 4.4 (i) follows from the linear combination of the yield locus of the skew

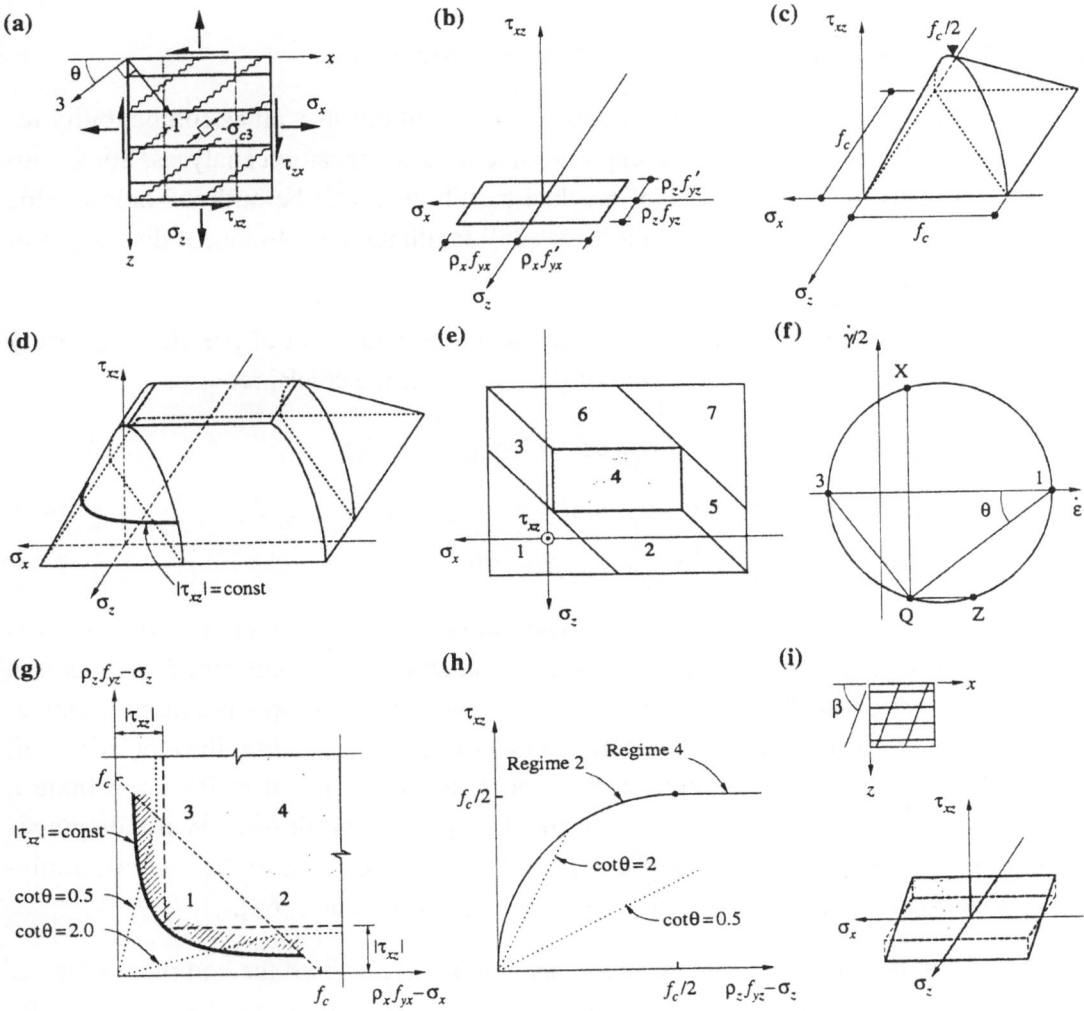

Fig. 4.4 – Limit analysis of membrane elements: (a) notation; (b) yield locus of reinforcement; (c) yield surface of plain concrete; (d) yield surface of reinforced concrete; (e) identification of yield regimes; (f) Mohr's circle of strain rates; (g) reinforcement design; (h) shear resistance governed by crushing of concrete; (i) yield locus of skew reinforcement.

reinforcement with that of the reinforcement in x-direction. The yield surface of the reinforced concrete element is again obtained geometrically by translating the yield surface of the plain concrete such that its origin moves within the yield locus of the reinforcement. The two cones of the yield surface of plain concrete will always remain intact in this procedure since the vector of resistance of any reinforcement layer is always parallel to a generatrix of the cones [82]. Multiple reinforcement layers and more sophisticated yield criteria for the concrete, accounting for the enhanced strength of concrete in biaxial compression or its tensile strength, have also been applied [109,81,78].

4.3 Compression Field Approaches

4.3.1 Stresses and Strains in Cracked Concrete Membranes

Compression field approaches satisfy equilibrium conditions as well as compatibility requirements and allow carrying out comprehensive load-deformation analyses. Some fundamental aspects of the behaviour of cracked membranes will be outlined below. This general examination of the inherent problems will facilitate the subsequent discussion of compression field approaches.

Consider an orthogonally reinforced concrete panel with a set of parallel, uniformly spaced cracks, Fig. 4.5 (a). Equilibrium of the stresses at the cracks requires

$$\sigma_x = \rho_x \sigma_{sxr} + \sigma_{cnr} \sin^2 \theta_r + \sigma_{ctr} \cos^2 \theta_r - \tau_{ctnr} \sin(2\theta_r)$$

$$\sigma_z = \rho_z \sigma_{szr} + \sigma_{cnr} \cos^2 \theta_r + \sigma_{ctr} \sin^2 \theta_r + \tau_{ctnr} \sin(2\theta_r) \qquad (4.9)$$

$$\tau_{xz} = (\sigma_{cnr} - \sigma_{ctr}) \sin\theta_r \cos\theta_r - \tau_{ctnr} \cos(2\theta_r)$$

see Fig. 4.5 (b). Introducing coordinates n and t aligned with the crack direction, replacing steel and bond stresses by equivalent stresses resulting from a uniform distribution in the transverse direction between the individual reinforcing bars, and assuming homogeneous material properties it can be concluded that the displacements at the cracks as well as the strains in the concrete between the cracks are independent of the coordinate t, Fig. 4.5 (c). Thus, since $\varepsilon_n = \partial u / \partial n$, $\varepsilon_t = \partial v / \partial t$ and $\gamma_{nt} = \partial u / \partial t + \partial v / \partial n$ the displacement u in n-direction is a function of n only and $\partial \gamma_{nt} / \partial t = 0$ leads to $\partial \varepsilon_t / \partial n = 0$, implying $\varepsilon_t = $ constant, where $v = $ concrete displacement in t-direction [67].

If the crack inclination θ_r and the crack spacing s_{rm} are known the complete states of stress and strain can be determined for given constitutive equations, i.e. stress-strain relationships for the concrete and the reihforcing steel as well as a bond shear stress-slip relationship and an aggregate interlock relationship $\sigma_{cnr} = \sigma_{cnr}(\delta_n, \delta_t)$, $\tau_{ctnr} = \tau_{ctnr}(\delta_n, \delta_t)$. Consideration of the conditions along the n-axis ($t = 0$) is sufficient since all relevant functions are independent of t. A suitable algorithm involving the steel stresses σ_{sxr} and σ_{szr} and the concrete stress component σ_{ctr} at the cracks ($n = s_{rm}/2$), the crack opening

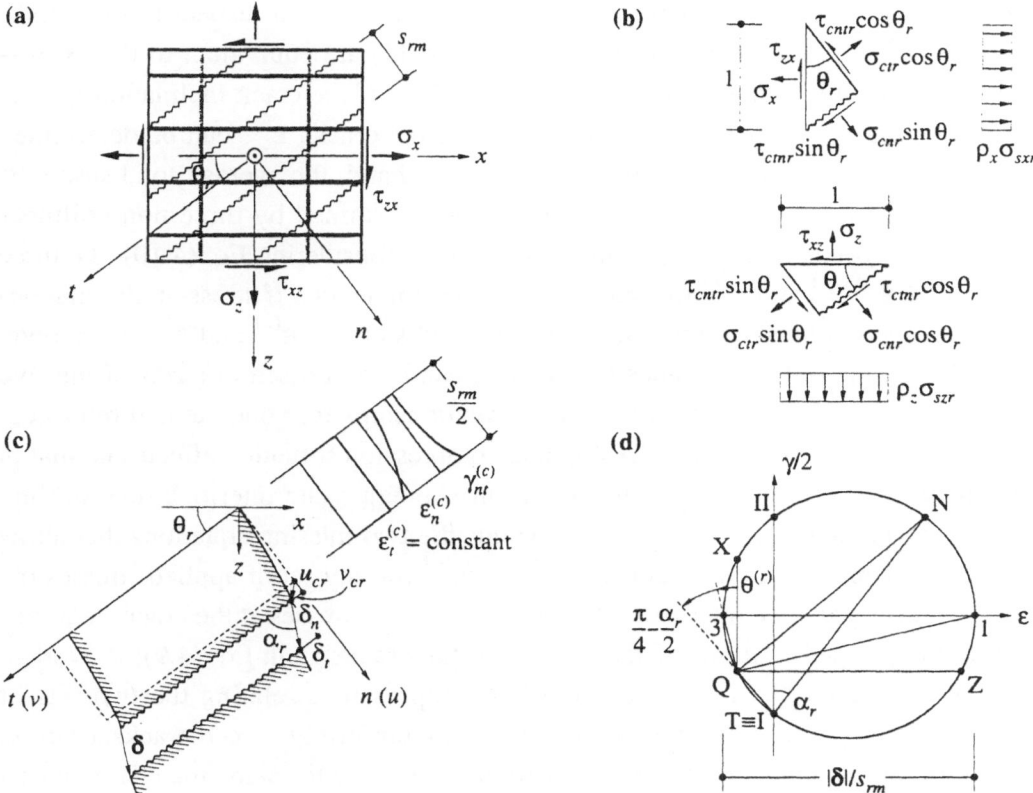

Fig. 4.5 – Cracked membrane: (a) notation; (b) stresses at crack; (c) displacements and strains; (d) Mohr's circle of average strains due to cracks.

δ_n and the crack slip δ_t as well as the concrete displacement components u_{cr} and v_{cr} at the cracks as the primary unknowns has been given in [67]. The algorithm could be implemented in a computer program; however, the solution is numerically intricate. The procedure presented in [67] is the most general possible approach provided that only one set of cracks and uniform distributions of steel and bond shear stresses in the transverse direction between the individual reinforcing bars are considered.

Neglecting local variations of the strains, a less general approach involving average strains can be established. The average total strains $\boldsymbol{\varepsilon}$ of a panel can be subdivided into average strains due to deformation of the concrete between the cracks $\boldsymbol{\varepsilon}^{(c)}$ and average strains $\boldsymbol{\varepsilon}^{(r)}$ due to the crack displacements

$$\boldsymbol{\varepsilon} = \boldsymbol{\varepsilon}^{(c)} + \boldsymbol{\varepsilon}^{(r)} \tag{4.10}$$

Considering a crack displacement inclined at an angle α_r with respect to the crack, Fig. 4.5 (c), the average strains due to cracks $\boldsymbol{\varepsilon}^{(r)}$ are equal to $\varepsilon_n^{(r)} = \delta_n / s_{rm}$, $\varepsilon_t^{(r)} = 0$ and $\gamma_{nt}^{(r)} = \delta_t / s_{rm}$. From Mohr's circle of strains, Fig. 4.5 (d), one obtains

$$\varepsilon_1^{(r)} = \frac{|\delta|}{2 s_{rm}}(1 + \sin \alpha_r) \qquad \varepsilon_3^{(r)} = -\frac{|\delta|}{2 s_{rm}}(1 - \sin \alpha_r) \qquad \theta^{(r)} = \theta_r + \frac{\alpha_r}{2} - \frac{\pi}{4} \tag{4.11}$$

for the principal strains due to cracks and their inclination with respect to the x-axis, where $|\boldsymbol{\delta}| = \sqrt{\delta_n^2 + \delta_t^2}$ and $\tan\alpha_r = \delta_n/\delta_t$. The Eqs. (4.11) are equivalent to the expressions obtained for the strains in a failure line, Eq. (3.9). If the crack inclination θ_r and the crack spacing s_{rm} are known, the strains due to the cracks $\boldsymbol{\varepsilon}^{(r)}$ can be determined from the crack displacements δ_n and δ_t. On the other hand, the average total strains $\boldsymbol{\varepsilon}$ and the concrete strains $\boldsymbol{\varepsilon}^{(c)}$ are each completely determined by three non-collinear strains since the y-axis is a principal direction. Hence, considering Eq. (4.10), the three states of strain $\boldsymbol{\varepsilon}$, $\boldsymbol{\varepsilon}^{(c)}$ and $\boldsymbol{\varepsilon}^{(r)}$ together contain five unknowns. If stress-strain relationships relating the steel and concrete stresses at the cracks to $\boldsymbol{\varepsilon}$, $\boldsymbol{\varepsilon}^{(c)}$ and $\boldsymbol{\varepsilon}^{(r)}$ are known, all quantities on the right-hand sides of Eq. (4.9) can be expressed in terms of the five primary unknowns. Note that the stress-strain relationships for concrete and reinforcement have to account for tension stiffening and compression softening effects and that in general, they depend on all three states of strain. An aggregate interlock relationship, $\sigma_{cnr} = \sigma_{cnr}(\delta_n, \delta_t)$ and $\tau_{ctnr} = \tau_{ctnr}(\delta_n, \delta_t)$, yields the two missing equations that allow determining the three states of strain $\boldsymbol{\varepsilon}$, $\boldsymbol{\varepsilon}^{(c)}$ and $\boldsymbol{\varepsilon}^{(r)}$ for any set of applied stresses σ_x, σ_z and τ_{xz}. Since aggregate interlock relationships involve stresses at the cracks, equilibrium is best expressed in terms of the stresses at the cracks as in Eq. (4.9); if average stresses between the cracks are used, additional expressions relating the two sets of stresses have to be postulated. Several sets of parallel, uniformly spaced cracks, each set having a different inclination, can be dealt with in essentially the same manner. Each set of cracks yields two additional unknowns (the strains due to cracks) and two additional equations (aggregate interlock relationship). However, reliable stress-strain relationships accounting for several sets of cracks are very difficult to establish.

For unbonded reinforcement, the procedure outlined above is particularly simple and equivalent to the general approach presented in [67]. Steel and concrete stresses are constant within the entire panel; the steel stresses follow from the total strains and the stress-strain relationship of the naked reinforcing bars, and the concrete stresses can be determined from the concrete strains and a biaxial stress-strain relationship of plain concrete. The principal compressive direction $\theta^{(c)}$ of the average concrete strains coincides with the principal compressive direction of the concrete stresses at the cracks; this condition is not exactly satisfied for bonded reinforcement.

According to the third Eq. (4.11), the crack direction t coincides with the principal compressive direction of the average strains due to the cracks if and only if the cracks open orthogonally, $\delta_t = 0$. On the other hand, the crack direction t coincides with the principal compressive direction of the concrete stresses if and only if no shear stresses are transferred across the cracks, $\tau_{ctnr} = 0$. Hence, even if the principal compressive direction $\theta^{(c)}$ of the average concrete strains is assumed to coincide with the principal compressive direction of the concrete stresses at the cracks, the principal direction θ of the total strains will neither coincide with $\theta^{(c)}$, $\theta^{(r)}$, nor θ_r unless the cracks are stress-free and open perpendicularly to their direction. In the latter case, the solution procedure outlined above fails since the strains due to cracks $\boldsymbol{\varepsilon}^{(r)}$ cannot be determined from the aggregate interlock relationship (stresses at the cracks $\tau_{ctnr} = 0$ and $\sigma_{cnr} = 0$ correspond to crack displacements $\delta_t = 0$ and $\delta_n > 0$, i.e., only the direction of the crack opening is known, not its amount). A suitable solution procedure for such cases is given below.

Compression field approaches that consider interlocked cracks of fixed direction rely on aggregate interlock relationships which, as stated in Chapter 2.2.5, depend on several parameters and are per se much less reliable than stress-strain relationships for steel or concrete. In addition, for given strains due to cracks, the crack displacements δ_n and δ_t are in direct proportion to the crack spacing which is subject to wide scatter, while aggregate interlock relationships are highly non-linear and sensitive to very small deformations. Furthermore, deformed reinforcing bars tend to expand the surrounding concrete, Chapter 2.4.1, resulting in smaller crack widths; this effect is very difficult to quantify. In light of all these problems, it can be concluded that the establishment of a compression field approach considering fixed, interlocked cracks suitable for design purposes is virtually impossible. In research and analysis situations, where the crack inclinations and the crack spacings are known beforehand, such models may nevertheless yield valuable insight into the actual behaviour of a structure.

Rather than fixed, interlocked cracks, most compression field approaches consider fictitious, rotating and orthogonally opening cracks that are stress-free and parallel to the principal compressive direction of concrete stresses. The principal compressive direction of the concrete strains is assumed to coincide with the principal compressive direction of the concrete stresses, $\theta^{(c)} = \theta^{(r)} = \theta_r$, and hence, the principal compressive direction of the total strains will also be equal to the crack inclination, $\theta = \theta_r$. Strictly speaking, stress-free cracks imply that the crack direction coincides with the principal compressive direction of the concrete stresses at the cracks; on the other hand, the principal compressive direction of the concrete strains is more closely approximated by the principal compressive direction of the average concrete stresses between the cracks than by the principal compressive direction of the concrete stresses at the cracks. Hence, if either only concrete stresses at the cracks or average stresses in the concrete between the cracks are considered, one of the assumed collinearities will only approximately be satisfied, except for unbonded reinforcement or cracks with an infinitely narrow spacing.

In models involving rotating, stress-free cracks that open perpendicularly to their direction, the components of the total strains $\boldsymbol{\varepsilon}$ are typically chosen as the three primary unknowns. The strains due to cracks $\boldsymbol{\varepsilon}^{(r)}$ cannot be determined from an aggregate interlock relationship; however, they can be eliminated from the calculations as long as the stresses in the concrete and in the reinforcement do not depend on the principal tensile concrete strain $\varepsilon_1^{(c)}$. Due to the assumptions made, the crack displacements create only strains in the principal tensile direction, $\varepsilon_3^{(r)} = 0$, and hence, $\varepsilon_3^{(c)} = \varepsilon_3$. Thus, as long as the stresses in the concrete and in the reinforcement depend only on $\varepsilon_3 = \varepsilon_3^{(c)}$ and $\varepsilon_1 = \varepsilon_1^{(c)} + \varepsilon_1^{(r)}$, but not on the principal tensile concrete strain $\varepsilon_1^{(c)}$ alone, all quantities on the right sides of Eq. (4.9) can be expressed in terms of the three remaining unknowns $\boldsymbol{\varepsilon}$, noting that $\tau_{ctnr} = \sigma_{cnr} = 0$. If the stresses in the concrete or in the reinforcement depend on $\varepsilon_1^{(c)}$, a solution is impossible.

If complete load-deformation analyses rather than the deformations corresponding to a given set of applied stresses have to determined, it is advisable to perform the calculations by incrementing ε_3 (or any other strain) rather than τ_{xz}. Since Eq. $(4.9)_{1,2}$ are independent of τ_{xz}, the two missing strains can be determined from these conditions alone for each set of σ_x, σ_z and ε_3, and τ_{xz} is obtained from the third Eq. (4.9) after the iterations. Incrementing ε_3 rather than τ_{xz} improves the convergence since only two non-linear equations have to be solved and it avoids numerical difficulties resulting from strain-softening branches of the stress-strain relationship of the concrete.

Several sets of parallel, uniformly spaced cracks, each set having a different inclination, and one set of rotating, stress-free cracks parallel to the principal compressive direction of concrete stresses can be dealt with in essentially the same manner, as long as the stresses in the concrete and in the reinforcement do not depend on $\varepsilon_1^{(c)}$ alone, but only on $\varepsilon_1 = \varepsilon_1^{(c)} + \varepsilon_1^{(r)}$.

Compression field approaches for regions of girders where all static and geometric quantities vary only gradually along the beam axis can be similarly developed. Equilibrium in the axial direction is satisfied integrally over the entire cross-section by determining the chord forces in essentially the same way as in the sectional design approach, Eq. $(4.3)_1$. By assuming a certain variation of the axial strains over the depth of the cross-section, the axial strains in the web are determined by the chord deformations and hence, the equilibrium condition in the axial direction is not involved in the calculation of the states of stress and strain in the web. Generally, the chord forces have to be adapted iteratively, depending on the axial stresses in the web obtained from the assumed variation of the axial strains; the resultant of the horizontal components of the diagonal concrete compressive stresses in the web is not necessarily acting at mid-depth of the web since the principal compressive stress trajectories may be curved. Implicitly, one assumes that the stresses in the web are equilibrated by the adjoining sections; if no horizontal web reinforcement is provided, the end-sections of the girder have to resist the resulting axial stresses. The calculations are often performed at mid-depth of the girder as if the stresses and strains were constant over the depth of the cross-section. However, even for a constant inclination of the principal compressive direction linearly varying transverse and shear strains result from a linear variation of the axial strains [109,83]; further details are given in Chapter 6.2.

4.3.2 Original Compression Field Approaches

Original compression field approaches consider fictitious rotating cracks with a vanishing spacing that are stress-free, parallel to the principal compressive direction of concrete stresses, and whose opening is restricted to be perpendicular to their direction. Using the notation of Chapter 4.3.1, $\theta^{(r)} = \theta^{(c)} = \theta = \theta_r$ is assumed and local variations of the stresses in the concrete and in the reinforcement due to bond action are neglected. Thus, a uniform uniaxial compressive stress field exists in the concrete. The principal direc-

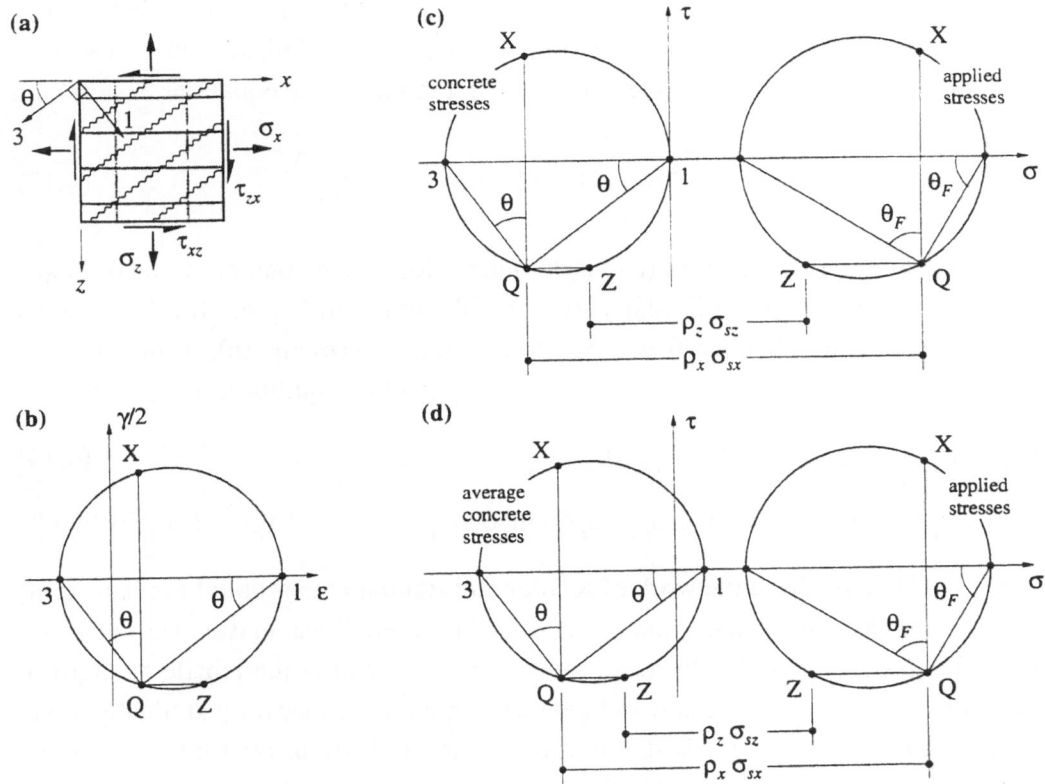

Fig. 4.6 – Compression field approaches: (a) notation; (b) strains; (c) stresses according to original compression field approach; (d) stresses according to Vecchio and Collins' modified compression field approach [156].

tions are free to adapt as required by the applied loads during the loading process; generally, they will rotate with increasing load.

The first application of a compression field approach is due to Kupfer [72] who supplemented Mörsch's [106] truss model by an equation for the inclination of the diagonal cracks in the web of a reinforced concrete girder. Kupfer analysed a linear elastic truss model, neglecting the tensile strength of concrete, and found his equation by minimising the complementary strain energy [47,165] of the truss with respect to the crack inclination θ. The same equation is obtained by expressing strain compatibility at mid-depth of the web, i.e., $\cot^2\theta = (\varepsilon_z - \varepsilon_3)/(\varepsilon_x - \varepsilon_3)$, see Fig. 4.6 (b), where $\varepsilon_x = (\varepsilon_{x\,sup} + \varepsilon_{x\,inf})/2$, $\varepsilon_z = \sigma_{sz}/E_s = \tau_{xz}\tan\theta/(\rho_z E_s)$ and $\varepsilon_3 = -\tau_{xz}(\tan\theta + \cot\theta)/E_c$, see Eq. (4.3). Kupfer's original equation can be expressed as

$$\tan^2\theta(1 + n\rho_z) = \cot^2\theta\, n\rho_z + \cot\theta\frac{\rho_z E_s(\varepsilon_{x\,sup} + \varepsilon_{x\,inf})}{2\tau_{xz}} \qquad (4.12)$$

where $n = E_s/E_c$ = modular ratio and $\varepsilon_{x\,sup}$, $\varepsilon_{x\,inf}$ are the axial strains of the top and bottom chord, respectively. Baumann [10] extended Kupfer's work to orthogonally reinforced membrane elements in plane stress. While he included shear forces acting on the crack faces in his general analysis, he did not take into account the accompanying normal forces; the results of his general investigation are therefore of limited value. However, for rotating, stress-free cracks, Baumann was the first to derive the equation

$$\tan^2\theta\rho_x(1+n\rho_z) + \tan\theta\rho_x\frac{\sigma_z}{\tau_{xz}} = \cot^2\theta\rho_z(1+n\rho_x) + \cot\theta\rho_z\frac{\sigma_x}{\tau_{xz}} \tag{4.13}$$

for the principal compressive stress (and strain) direction in the concrete of an orthogonally reinforced concrete panel in plane stress as illustrated in Fig. 4.6 (a). Baumann's equation can be directly obtained from considering Figs. 4.6 (b) and (c), using the compatibility condition $\cot^2\theta = (\varepsilon_z - \varepsilon_3)/(\varepsilon_x - \varepsilon_3)$ and noting that equilibrium requires

$$\sigma_x = \sigma_{cx} + \rho_x\sigma_{sx} \qquad \sigma_z = \sigma_{cz} + \rho_z\sigma_{sz} \qquad \tau_{xz} = \tau_{cxz} \tag{4.14}$$

where $\sigma_{cx} = E_c\varepsilon_3\cos^2\theta$, $\sigma_{sx} = E_s\varepsilon_x$, $\sigma_{cz} = E_c\varepsilon_3\sin^2\theta$, $\sigma_{sz} = E_s\varepsilon_z$ and $\tau_{cxz} = -E_c\varepsilon_3\sin\theta\cos\theta$.

Potucek [121] generalised the work of Kupfer and Baumann to account for skew reinforcement meshes and investigated the behaviour of cracked, linear elastic webs of structural concrete girders subjected to shear and flexure, neglecting the tensile strength of concrete. For regions where all static and geometric quantities vary only gradually along the girder axis, he assumed a linear distribution of the axial strains over the depth of the cross-section and obtained curved trajectories of the principal compressive stresses in the web. This is in accordance with the initial orientation of the diagonal shear cracks in the web of girders observed in tests. In Chapter 6.2, a generalisation of this approach to nonlinear material behaviour will be presented. For the analysis of the support regions, Potucek developed a computer program based on the finite element method. Thürlimann and Lüchinger [151] used a space truss model similar to Kupfer's approach to compute the deformations of beams subjected to torsion and flexure.

While Kupfer's and Baumann's work constitutes a linear compression field approach, Collins and Mitchell independently developed a similar, non-linear approach for torsion [103] and shear [27] problems and presented a comprehensive paper on the analysis and the design of prestressed and non-prestressed concrete beams subjected to combined actions [29]. Starting from the observation that the behaviour of webs of concrete girders without tensile strength is similar to the post-buckling behaviour of thin-webbed metal girders in shear [160], they obtained the condition $\cot^2\theta = (\varepsilon_z - \varepsilon_3)/(\varepsilon_x - \varepsilon_3)$ for the inclination of the compression diagonals with respect to the x-axis. This condition merely expresses strain-compatibility and is directly obtained from Mohr's circle of strains, Fig. 4.6 (b). In order to account for the experimentally observed softening of diagonally cracked concrete, Chapter 2.4.3, Collins [28] and Vecchio and Collins [154,155] proposed various expressions, leading to Eq. (2.25) for the peak concrete compressive stress, i.e. $f_c / f_c' = (0.8 + 170\varepsilon_1)^{-1} \le f_c'$, where $\varepsilon_1 = \varepsilon_x + (\varepsilon_x + 0.002)\cot^2\theta$ and $\varepsilon_x =$ axial strain at mid-depth of the web of a member, to be determined from a strain-compatibility analysis for the section under the action of the moment M and the axial force $N + V\cot\theta$ (N and V = applied axial and shear forces).

Using the non-linear compression field approach developed by Collins et al., arbitrary stress-strain relationships for the concrete and the reinforcement can be dealt with. The approach was adopted by many other researchers and it was extended to prestressed beams and combined actions by Pré [124] and Teutsch and Kordina [150].

Original compression field approaches are directly related to limit analysis methods and allow carrying out complete load-deformation analyses of structural concrete members. However, response predictions typically overestimate the deformations since tension stiffening effects are completely neglected and rotating, stress-free rather than fixed, interlocked cracks are considered. If failure is governed by crushing of the concrete, the ultimate load may be seriously overestimated unless adequately softened stress-strain relationships for the concrete are adopted. Also, crack spacings and crack widths as well as the localisation of the steel strains at the cracks cannot be determined from original compression field approaches. Various modifications of the original compression field approaches have been established in order to overcome these difficulties.

4.3.3 Modified Compression Field Approaches

The modifications of the original compression field approaches can be subdivided into approaches considering fixed, interlocked cracks and approaches accounting for tension stiffening effects through empirical relationships between average strains and average stresses between the cracks. Both types of modification yield stiffer response predictions than the original compression field approaches.

Vecchio and Collins [156] assumed that the principal directions of the average total strains and of the average concrete stresses between the cracks coincide and introduced empirical relationships between the average total strains and the average tensile and compressive stresses in the concrete between the cracks; these relationships were calibrated based on many tests on orthogonally reinforced panels [155,68,70,13]. For the reinforcement, they adopted the stress-strain relationship of naked steel bars without modification. By expressing equilibrium in terms of average stresses between the cracks, tension stiffening is implicitly taken into account, see Chapter 2.4.2. The average principal concrete compressive stress between the cracks σ_{c3m} is accompanied by an average principal tensile stress σ_{c1m}, and the quantities σ_{cx}, σ_{cz} and τ_{cxz} in Eq. (4.14) are equal to $\sigma_{c1m}\sin^2\theta + \sigma_{c3m}\cos^2\theta$, $\sigma_{c1m}\cos^2\theta + \sigma_{c3m}\sin^2\theta$ and $(\sigma_{c1m}-\sigma_{c3m})\sin\theta\cos\theta$, respectively, see Fig. 4.6 (d). A suitable solution procedure involving the average total strains as the three primary unknowns is outlined in Chapter 4.3.1. Vecchio and Collins' [156] modified compression field approach substantially contributed to the present knowledge in the field of membrane shear behaviour; in fact, it was the first model accounting for tension stiffening in reinforced concrete membranes. However, the model has one major drawback stemming from the use of the stress-strain relationship of the naked steel bars in terms of average stresses and average strains. As shown in Chapter 2.4.2, equilibrium requires that the tensile stresses in the concrete and in the reinforcement are related to each other by Eq. (2.24). The use of the stress-strain relationship of naked steel in terms of average stresses and strains corresponds to replacing the average steel stresses σ_{sm} on the left hand side of Eq. (2.24) by the maximum steel stresses at the cracks σ_{sr}, resulting in yield and ultimate stresses of the reinforced concrete tension chord exceeding those of the naked steel by $\sigma_{cm}(1-\rho)/\rho$. This difference results in an overestimation of the yield and ultimate loads, particularly for small reinforcement ratios. In order to overcome this discrepancy, Vecchio and Collins [156] introduced an additional check of the stresses at the cracks, allowing for substantial shear stresses acting on the crack faces. However, this is incompatible with the basic assumption of coinciding principal directions of the average total strains and the average concrete stresses, which implies cracks that are parallel to the principal compressive direction of the average concrete stresses and whose opening is restricted to be perpendicular to their direction, Chapter 4.3.1.

The so-called rotating-angle softened truss model developed by Hsu [57] is essentially the same as Vecchio and Collins' [156] modified compression field approach, eliminating the discrepancy outlined above by adopting a suitably reduced relationship between average stresses and average strains in the reinforcement. Still, some other difficulties are inherent to approaches expressing equilibrium in terms of average stresses in the concrete between the cracks; further details are given in Chapter 5.3.1.

Assuming a uniform, uniaxial compressive stress field in the web concrete, Kupfer et al. [74] presented a compression field approach for webs of girders with interlocked cracks of fixed direction. They subdivided the total strains $\boldsymbol{\varepsilon}$ into concrete strains $\boldsymbol{\varepsilon}^{(c)}$ and strains due to cracks $\boldsymbol{\varepsilon}^{(r)}$ as outlined in Chapter 4.3.1, see Eq. (4.10). They neglected possible tensile stresses σ_{c1} in the concrete resulting from aggregate interlock action, for which they adopted a linear fit of Walraven's [161] experiments, and a semi-empirical crack spacing formula established for uniaxial tension. For the concrete, Kupfer et al. used a parabolic uniaxial stress-strain curve, reducing the concrete strength by 15% as proposed by Robinson and Demorieux [130] in order to account for compression softening effects. They considered cracks inclined at 45 degrees to the girder axis, vertical stirrups and average values of the longitudinal strains and the crack displacements at mid-depth of the web, performing the calculations for two distinct values of the axial strains, i.e., $\varepsilon_x = 0$ and $\varepsilon_x^{(r)} = 0$. They assumed that the stirrups start to yield at the ultimate state, $\varepsilon_{sz} = f_{syz}/E_s$, implicitly treating unbonded reinforcement while accounting for shrinkage strains and anchorage slip of the stirrups by reducing ε_{sz} as compared to ε_z. In such approaches, the solution procedure outlined in Chapter 4.3.1 can be simplified since the equilibrium conditions $(4.9)_{1,2}$ are replaced by the assumption of ε_x and ε_z and hence, there remain three rather than five unknowns and equations. Kupfer et al. [74] pointed out that the principal compressive direction θ of the average total strains will generally deviate from the principal compressive direction of the concrete stresses (and strains) unless the cracks open orthogonally.

Reducing the vertical strains by a constant factor of 0.8, Kirmair and Mang [69] included tension stiffening effects in Kupfer et al.'s [74] model while still assuming

$\varepsilon_{sz} = f_{sy}/E_s$ at the ultimate state. They considered a biaxial state of stress in the concrete between the cracks, adopting a biaxial stress-strain relationship [73], and determined the crack inclination from the principal stress direction in the uncracked state, including axial forces and prestressing. They adopted an improved fit of Walraven's aggregate interlock relationship [162], accounting separately for shear and normal stresses acting on the cracks, and assumed a constant value of 200 mm for the crack spacing measured along the stirrups. Kirmair and Mang considered vertical stirrups and axial strains in the range $\varepsilon_x = -2\%o$ to $2\%o$ and they pointed out that the amount by which the shear resistance can be increased beyond the resistance according to the 45-degree truss model depends on the axial strains. Kupfer and Bulicek [75] supplemented the model by a more realistic tension stiffening factor depending on the stirrup reinforcement ratio, still assuming $\varepsilon_{sz} = f_{sy}/E_s$ at the ultimate state, and used a biaxial concrete stress-strain relationship accounting for the influence of the concrete strength [35]. They demonstrated that generally, the concrete between the cracks may be subjected to biaxial compression or compression-tension, but that the lateral compressive or tensile stresses are small and can be neglected in design practice. They re-emphasised the strong dependence of the shear resistance on the axial strains and claimed that for initial cracks inclined at 45 degrees, no flatter inclination of the truss model than 45 degrees would be possible for axial strains larger than about $\varepsilon_x \approx 1\%o$, and that even for values of $\varepsilon_x \approx -1\%o$, the shear resistance would be significantly lower than according to the variable angle truss model of limit analysis due to interface shear failures along the cracks of fixed direction.

Reineck and Hardjasaputra [126] obtained similar results from their model; they considered a uniaxial compressive stress field in the concrete between the cracks, also assuming $\varepsilon_{sz} = f_{sy}/E_s$ at the ultimate state. Rather than adopting an aggregate interlock relationship, they assumed that the crack opening is perpendicular to the principal direction of the concrete stresses (and strains), i.e., that the latter is a characteristic direction of the strains due to the cracks $\boldsymbol{\varepsilon}^{(r)}$ and hence, no axial strains due to cracks are obtained in the principal compressive direction of the concrete stresses (and strains). This allows one to use a more direct solution procedure similar to that adopted for models involving rotating, stress-free cracks that open orthogonally, see Chapter 4.3.1.

Bulicek and Kupfer [18,19] adopted Walraven's [162] full aggregate interlock relationship and determined the crack spacing from the condition $\sigma_{c1} = f_{ct}$ between the cracks. Furthermore, they optimised the value of the stirrup strains $\varepsilon_{sz} \geq f_{sy}/E_s$ in order to obtain the maximum shear resistance rather than assuming $\varepsilon_{sz} = f_{sy}/E_s$ as in the previous investigations [74,69,126,75]. With this more realistic approach, they obtained a much less pronounced dependence of the shear resistance on the axial strains and only small reductions with respect to limit analysis, even for axial strains as high as $\varepsilon_x \approx 2\%o$. Taking into account that crack spacings and aggregate interlock relationships are subject to considerable scatter, Chapter 4.3.1, it appears that the shear resistance determined according to limit analysis, using a suitably reduced value of the effective concrete compressive strength, does not have to be reduced due to interface shear problems.

Dei Poli et al. [39,40] developed a model similar to that of Kupfer et al. [74], adopting Bazant and Gambarova's rough crack model [11] for aggregate interlock and accounting for tension stiffening effects by a parameter evaluated from bond shear stress-slip relationships. Carrying out the calculations for the same values of the axial strains as in [74], they included eccentric compression of the concrete struts, assuming that the latter fail in shear-compression along the compression chord. While Dei Poli et al. [39,40] assumed $\varepsilon_{sz} = f_{sy}/E_s$ at the ultimate state, di Prisco and Gambarova [125] accounted for plastic stirrup strains $\varepsilon_{sz} \geq f_{sy}/E_s$ similar to Bulicek and Kupfer [18,19] and considered non-uniform crack displacements as well as an empirical contribution of dowel action.

Pang and Hsu [118] considered orthogonally reinforced concrete panels with interlocked cracks of fixed direction. Rather than adopting an aggregate interlock relationship, they neglected the normal stresses acting on the crack faces and introduced an empirical relationship between the shear stresses acting on the crack faces and the associated average shear strains due to crack slip, $\tau_{ctnr} = \tau_{ctnr}(\gamma_{nt}^{(r)})$. They expressed equilibrium in terms of average stresses between the cracks and adopted essentially the same empirical relationships between average stresses and average strains for concrete and reinforcement as in the so-called rotating angle softened truss model [57].

4.4 Finite Element Methods

In design practice, the state of stress in a structure is often obtained from a linear elastic finite element analysis. Based on this statically admissible state of stress, the reinforcement can be designed using limit analysis methods, Eq. (4.8). Although this dimensioning procedure can be justified by the lower-bound theorem of limit analysis, Chapter 3.1, it is not entirely satisfactory. While the stresses resulting from a linear elastic analysis typically require continuously varying amounts of reinforcement and often exhibit sharp peaks at static or geometric discontinuities, the designer is faced with discrete bar diameters and spacings and would prefer to provide a concentrated main reinforcement and a uniform, well-distributed minimum reinforcement in the remaining parts of the structure.

Recently, finite element programs were developed that allow an elasto-plastic dimensioning based on the superposition of self-equilibrated stress-fields on the linear elastic solution [8]. In these programs, the designer specifies a minimum reinforcement and the locations where the main reinforcement has to be provided. The programs create and optimise self-equilibrated states of stress and dimension the main reinforcement, accounting for the minimum reinforcement. In this way, the designer uses engineering judgment in dimensioning the structure, and efficient reinforcement layouts can be obtained. Implicitly the programs assume sufficient deformation capacity of the structure; refined models like the cracked membrane model, Chapter 5, would allow performing non-linear finite element calculations in order to discuss the related questions of the demand and supply of deformation capacity.

5   Behaviour of Membrane Elements

5.1   General

In Chapter 4.3.1, basic aspects of stresses and strains in cracked reinforced concrete elements subjected to in-plane shear and normal stresses have been discussed and computational procedures that allow treating cracks as fixed and interlocked rather than rotating and stress-free have been proposed; these considerations will not be repeated here. The establishment of a general model accounting for fixed, interlocked cracks suitable for design purposes is considered to be impossible; owing to their sensitivity to crack spacings, aggregate interlock relationships are too unreliable. Furthermore, models considering fixed, interlocked cracks generally do not allow for a direct comparison with limit analysis methods, while one of the primary objectives of this thesis consists in supplementing the theory of plasticity by refined models on whose basis a discussion of the questions of the demand for and the supply of deformation capacity is possible. Hence, apart from the examination of the load-carrying behaviour of uniaxially reinforced elements presented in Chapter 5.4.3, approaches considering fixed, interlocked cracks are not further discussed. Rather, a model considering rotating, stress-free cracks is established.

The cracked membrane model [67] presented in this chapter is a new model for cracked, orthogonally reinforced concrete panels subjected to a homogeneous state of plane stress, combining the basic concepts of the original compression field approaches, Chapter 4.3.2, and the tension chord model, Chapter 2.4.2. Crack spacings and tensile stresses between the cracks are determined from first principles and the link to limit analysis is maintained since equilibrium conditions are expressed in terms of stresses at the cracks rather than average stresses between the cracks as in the modified compression field approach proposed by Vecchio and Collins [156], Chapter 4.3.3. Hence, while enhancing the stiffness, concrete tensile stresses do not directly influence the ultimate strength. Indirectly, there is a minor influence for failures governed by crushing of the concrete since ε_1 is somewhat reduced and the softening of concrete in compression is less pronounced, see Chapter 2.4.3.

Both a general numerical method and an approximate analytical solution are derived and the results are compared with previous theoretical and experimental work. The results presented in [67] are supplemented by expressions for crack widths, including the effect of Poisson's ratio, as well as a detailed comparison with limit analysis methods. In addition, high reinforcement ratios as well as diagonal crack spacings s_{rm} below the maximum value s_{rmo} are considered, i.e., $s_{rmo}/2 \leq s_{rm} \leq s_{rmo}$, see Chapter 2.4.2.

5.2 Cracked Membrane Model

5.2.1 Basic Assumptions

Crack faces are assumed to be stress-free, able to rotate, and perpendicular to the principal tensile direction of the average strains. Thus, θ_r is a variable rather than a fixed angle and $\sigma_{cnr} = \tau_{ctnr} = 0$ in Eq. (4.9). Hence, the principal compressive direction of the concrete stresses at the cracks coincides with that of the total strains, Figs. 5.1 (a) and (b). Due to variations of the principal direction of the concrete stresses between the cracks, the assumed coaxiality will only approximately be satisfied in reality; however, deviations are small since the concrete deformations contribute comparatively little to the overall strains. The equilibrium conditions, Eq. (4.9), simplify to

$$\sigma_x = \rho_x \sigma_{sxr} + \sigma_{c3r} \cos^2 \theta_r$$

$$\sigma_z = \rho_z \sigma_{szr} + \sigma_{c3r} \sin^2 \theta_r \qquad (5.1)$$

$$\tau_{xz} = -\sigma_{c3r} \sin \theta_r \cos \theta_r$$

where $\sigma_{c3r} = \sigma_{ctr}$ and $\sigma_{c1r} = \sigma_{cnr} = 0$. Rearranging Eq. (5.1) one obtains the following expressions for the steel and concrete stresses at the cracks

$$\sigma_{sxr} = (\sigma_x + \tau_{xz} \cot \theta_r)/\rho_x$$

$$\sigma_{szr} = (\sigma_z + \tau_{xz} \tan \theta_r)/\rho_z \qquad (5.2)$$

$$\sigma_{c3r} = -\tau_{xz}/\sin \theta_r \cos \theta_r = -\tau_{xz}(\cot \theta_r + \tan \theta_r)$$

see Fig. 5.1 (b). Note that the coordinate system has been oriented such that the x- and z-axes correspond to the reinforcement directions, Fig. 5.1 (c).

Steel and bond shear stresses are dealt with according to Figs. 2.12 (a) and (b), and the basic concepts of the tension chord model are extended to cracked panels. In particular, similar to Eqs. (2.19) to (2.21) and Fig. 2.12 (d), relationships between the maximum steel stresses σ_{sxr} and σ_{szr} and the average strains ε_x and ε_z are established.

A parabolic stress-strain relationship, Eq. (2.4), is used for the concrete compressive stress σ_{c3r} at the crack, and compression softening effects as well as the influence of f_c' on the peak compressive stress f_c are accounted for according to Eq. (2.27), i.e.

$$\sigma_{c3r} = f_c(\varepsilon_3^2 + 2\varepsilon_3\varepsilon_{co})/\varepsilon_{co}^2 \qquad (5.3)$$

and

$$f_c = \frac{(f_c')^{2/3}}{0.4 + 30\varepsilon_1} \leq f_c' \quad \text{in MPa} \qquad (5.4)$$

where ε_{co} is the concrete strain at the peak compressive stress and ε_1 and ε_3 are the principal values of the total strains. Eq. (5.4) will be further commented on in Chapter 5.3.1.

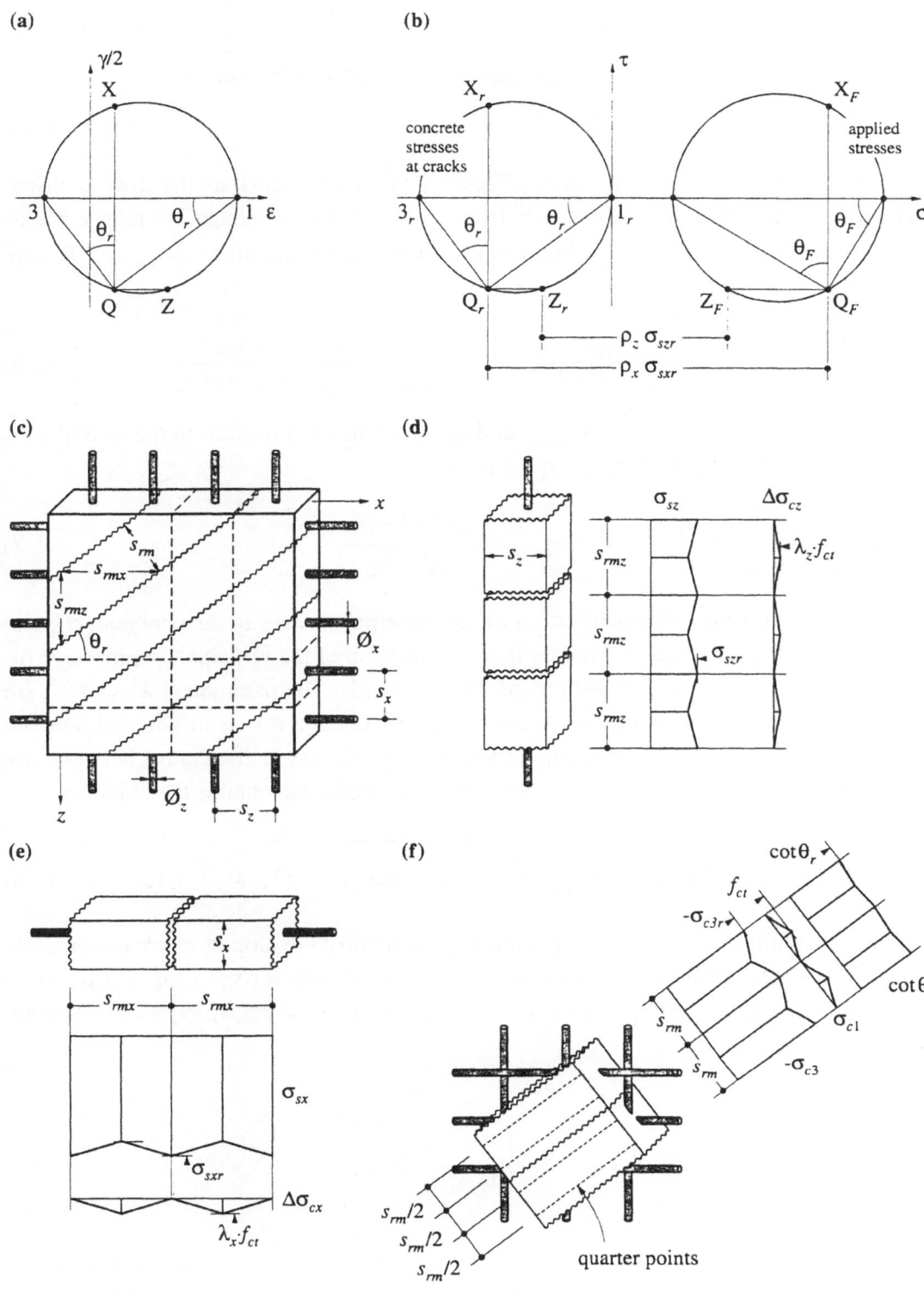

Fig. 5.1 – Cracked membrane model: (a) strains; (b) applied stresses and concrete stresses at the cracks; (c) bar and crack spacings; (d) vertical stresses; (e) horizontal stresses; (f) principal stresses in concrete, variation of principal concrete compressive stress direction and location of quarter points.

5.2.2 Crack Spacings and Concrete Stresses

According to Fig. 5.1 (c) the crack spacings are related to each other by

$$s_{rm} = s_{rmx}\sin\theta_r = s_{rmz}\cos\theta_r \qquad (5.5)$$

where s_{rm} is the diagonal crack spacing. Figs. 5.1 (d) and (e) illustrate the stress distribution between the cracks. At the centre between two consecutive cracks the tensile stresses transferred to the concrete by bond reach their maximum values $\Delta\sigma_{cx} = \lambda_x f_{ct}$ and $\Delta\sigma_{cz} = \lambda_z f_{ct}$, where

$$\lambda_x = \frac{\Delta\sigma_{cx}}{f_{ct}} = \frac{s_{rmx}}{s_{rmxo}} = \frac{s_{rm}}{s_{rmxo}\sin\theta_r} \qquad \lambda_z = \frac{\Delta\sigma_{cz}}{f_{ct}} = \frac{s_{rmz}}{s_{rmzo}} = \frac{s_{rm}}{s_{rmzo}\cos\theta_r} \qquad (5.6)$$

and the maximum crack spacings s_{rmxo} and s_{rmzo} for uniaxial tension in the x- and z-directions, respectively, follow from Eq. (2.18) as

$$s_{rmxo} = \frac{f_{ct}\varnothing_x(1-\rho_x)}{2\tau_{bo}}\frac{}{\rho_x} \qquad s_{rmzo} = \frac{f_{ct}\varnothing_z(1-\rho_z)}{2\tau_{bo}}\frac{}{\rho_z} \qquad (5.7)$$

Fig. 5.1 (f) illustrates the distribution of the principal stresses in the concrete, σ_{c1} and σ_{c3}, and the variation of $\cot\theta$, where θ is the inclination of the principal compressive direction of the concrete stresses with respect to the x-axis. The parameters λ_x and λ_z are no longer limited by the conditions $0.5 \le \lambda_x \le 1$ and $0.5 \le \lambda_z \le 1$ as in uniaxial tension. Rather, observing that the maximum concrete tensile stress at the centre between the cracks cannot be greater than the concrete tensile strength, one obtains the condition

$$\frac{f_{ct}}{2}(\lambda_x+\lambda_z) - \frac{\tau_{xz}}{2}(\cot\theta_r+\tan\theta_r) + \sqrt{\left[\frac{\tau_{xz}}{2}(\cot\theta_r-\tan\theta_r) - \frac{f_{ct}}{2}(\lambda_x-\lambda_z)\right]^2 + \tau_{xz}^2} \le f_{ct} \qquad (5.8)$$

for the maximum diagonal crack spacing s_{rmo} in a fully developed crack pattern, see Fig. 5.3 (a). At the limit, Eq. (5.8) can be solved numerically for s_{rmo}; Fig. 5.2 provides polar representations of the solution, assuming values of $\rho_x = 2.5\%$, $\varnothing_x = \varnothing_z = 16$ mm,

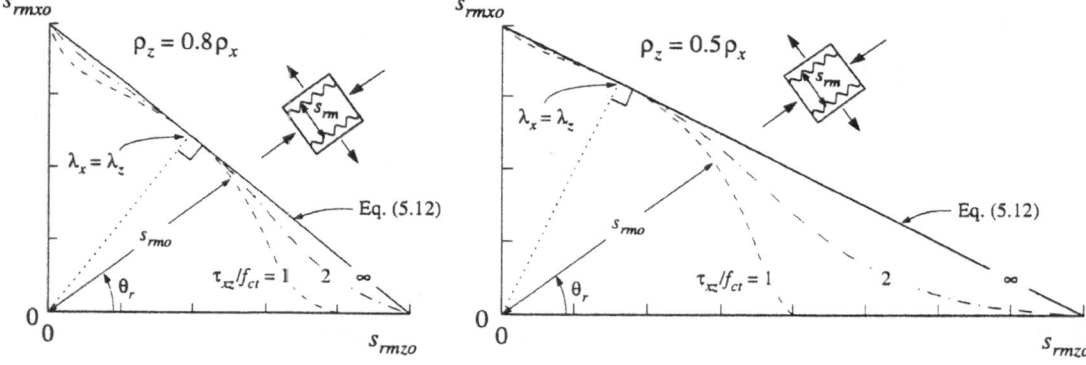

Fig. 5.2 – Cracked membrane model: polar representation of maximum diagonal crack spacing s_{rmo} for two combinations of reinforcement ratios.

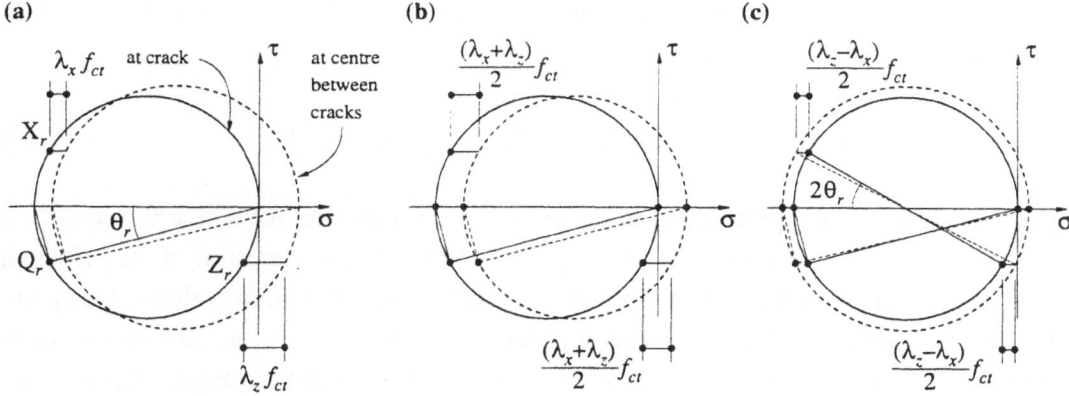

Fig. 5.3 – Cracked membrane model: (a) total stresses in the concrete at the cracks and at the centre between the cracks; (b), (c) subdivision of tensile stresses transferred to the concrete by bond into symmetric and antisymmetric portions.

$f_c' = 30$ MPa and $\tau_{bo}/2 = f_{ct} = 0.3(f_c')^{2/3}$ in MPa. If the cracks are almost parallel to the reinforcement resulting in the smaller crack spacing in uniaxial tension (x-reinforcement and s_{rmxo} in the examples), the principal compressive direction θ of the concrete stresses between the cracks varies significantly at small ratios τ_{xz}/f_{ct}. In particular, for $\tau_{xz}/f_{ct} = 1$, $s_{rmxo} < s_{rmzo}$ and $\theta_r \rightarrow 0$ one obtains $\cot(2\theta) = s_{rmxo}/(2s_{rmzo})$ at the centre between the cracks ($\theta = 34.1°$ and $\theta = 38.0°$ for the examples shown in Fig. 5.2). For $\tau_{xz}/f_{ct} = 1$, the maximum diagonal crack spacing s_{rmo} therefore equals s_{rmxo} (the smaller crack spacing in uniaxial tension) both for $\theta_r = \pi/2$ and $\theta_r = 0$, see Fig. 5.2. Crack inclinations $\theta_r \rightarrow 0$ are only obtained for $\rho_z \ll \rho_x$ or for predominant axial stresses $|\tau_{xz}| \ll |\sigma_z|$ or $|\tau_{xz}| \ll |\sigma_x|$, where $\sigma_x \ll \sigma_z$, i.e., if the principal tensile direction of the applied stresses is almost parallel to the (weaker) z-reinforcement. Hence, drastic variations of θ between the cracks rarely occur in practice. Also, for increasing ratios of τ_{xz}/f_{ct}, the variation of θ between the cracks diminishes and for $\tau_{xz}/f_{ct} \geq s_{rmzo}/s_{rmxo}$, one obtains s_{rmzo} for $\theta_r = 0$, as one would expect. For $s_{rmxo} > s_{rmzo}$ and crack inclinations close to $\theta_r = \pi/2$, similar remarks apply.

As in uniaxial tension, Chapter 2.4.2, the minimum crack spacing follows from the observation that a tensile stress equal to the concrete tensile strength must be transferred to the concrete in order to generate a new crack. Although the variation of the principal tensile stresses between the cracks is slightly non-linear according to Eq. (5.8), see Fig. 5.1 (f), the minimum diagonal crack spacing thus amounts to $s_{rmo}/2$ and the diagonal crack spacing s_{rm} in a fully developed crack pattern is limited by

$$s_{rmo}/2 \leq s_{rm} \leq s_{rmo} \tag{5.9}$$

or, equivalently, $0.5 \leq \lambda \leq 1$, where $\lambda = s_{rm}/s_{rmo}$.

An approximate solution for the maximum diagonal crack spacing s_{rmo} is obtained by subdividing the tensile stresses transferred to the concrete by bond into a symmetric and an antisymmetric portion, Figs. 5.3 (b) and (c). It can be seen that the maximum concrete tensile stress is approximately equal to

$$\frac{f_{ct}}{2}(\lambda_x + \lambda_z) - \frac{f_{ct}}{2}(\lambda_x - \lambda_z)\cos(2\theta_r) = f_{ct}(\lambda_x \sin^2\theta_r + \lambda_z \cos^2\theta_r) \tag{5.10}$$

Eq. (5.10) clarifies the approximation made in Figs. 5.3 (b) and (c); since θ_r = constant and λ_x as well as λ_z vary linearly with s_{rm}, Eq. (5.6), a linear variation of the principal tensile stresses in the concrete from zero at the cracks to the maximum value at the centre between the cracks, Eq. (5.10), has been assumed. Setting the expression on the right-hand side of Eq. (5.10) equal to λf_{ct}, where $0.5 \leq \lambda \leq 1$ according to Eq. (5.9), one gets

$$\lambda_x \sin^2\theta_r + \lambda_z \cos^2\theta_r = \lambda \tag{5.11}$$

which, for $\lambda = 1$, yields the approximation of the maximum diagonal crack spacing

$$s_{rmo} = \left(\frac{\sin\theta_r}{s_{rmxo}} + \frac{\cos\theta_r}{s_{rmzo}}\right)^{-1} \tag{5.12}$$

Eq. (5.12) corresponds to the expression for the diagonal crack spacing suggested by Vecchio and Collins (1986) without further justification. It coincides with the numerical solution both for $\lambda_x = \lambda_z$ and for high values of τ_{xz}/f_{ct} as shown in Fig. 5.2. In general, Eq. (5.12) provides an upper bound for the maximum diagonal crack spacing.

5.2.3 General Numerical Method

The solution procedure is the same as for the original compression field approaches, Chapter 4.3.1. Considering the average total strains ε_x, ε_z, and ε_3 as the primary unknowns and noting that $\varepsilon_1 = \varepsilon_x + \varepsilon_z - \varepsilon_3$ and $\cot^2\theta_r = (\varepsilon_z - \varepsilon_3)/(\varepsilon_x - \varepsilon_3)$ all quantities in Eq. (5.1) can be expressed as functions of these three unknowns. In fact, σ_{c3r} is determined by Eqs. (5.3) and (5.4) while σ_{sxr} and σ_{szr} follow from Eqs. (2.19) to (2.21), substituting appropriate values for f_{sy}, f_{su}, E_s, E_{sh}, τ_{bo}, τ_{b1}, ρ, $\varnothing$ and s_{rm} in the x- and z-direction, respectively, and noting that the crack spacings s_{rmx} and s_{rmz} are related by Eq. (5.5), while the diagonal crack spacing s_{rm} follows from Eq. (5.8) or its approximation (5.12). Thus, for any given set of applied stress components σ_x, σ_z and τ_{xz}, the strains ε_x, ε_z, and ε_3 can be determined from Eq. (5.1) in an iterative numerical manner.

If complete load-deformation responses rather than just deformations corresponding to a given set of applied stresses have to determined, it is advisable to perform the calculations by incrementing ε_3 rather than τ_{xz}. As pointed out in Chapter 4.3.1, this results in better convergence and avoids numerical difficulties resulting from strain-softening branches of the stress-strain relationship of the concrete. Fig. 5.4 illustrates the results of a typical load-deformation analysis, using the general expression (5.8) for the diagonal crack spacing and assuming $\lambda = 1$ and $\tau_{b1} = \tau_{bo}/2 = f_{ct} = 0.3(f_c')^{2/3}$ in MPa. Basic values underlying the calculations are indicated in Fig. 5.4.

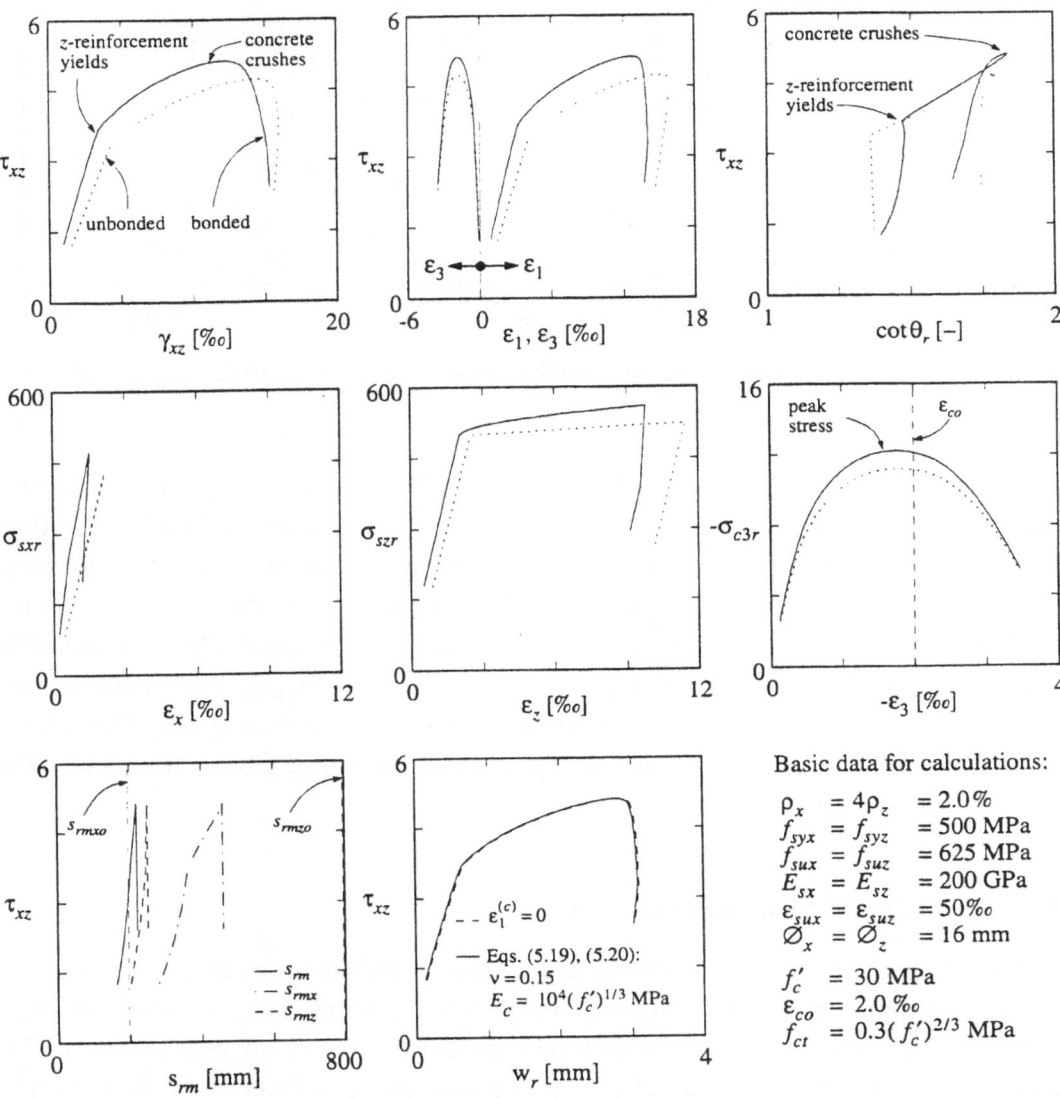

Fig. 5.4 – Cracked membrane model: typical load-deformation response according to the general numerical method. Note: stresses in MPa.

From Fig. 5.4 it can be seen that the cracked membrane model predicts failure by crushing of the concrete while the stronger *x*-reinforcement remains elastic. Tension stiffening results in a markedly stiffer response as can be recognized from the difference between the curves labelled "bonded" and "unbonded"; the ultimate load is also somewhat increased due to the reduction of ε_1, see Eq. (5.4). After onset of yielding of the weaker *z*-reinforcement, the deformations increase more rapidly and the crack direction rotates towards the *x*-axis until eventually, the ultimate load is reached. The concrete strain at the peak concrete compressive stress is somewhat lower than ε_{co} since near ε_{co}, the increase in the parabolic stress-strain curve of concrete, Eq. (5.3), is compensated by the progressive reduction of the compressive strength due to increasing ε_1, Eq. (5.4). A reduction of ε_{co} is also observed in tests [155]; partly, this phenomenon can

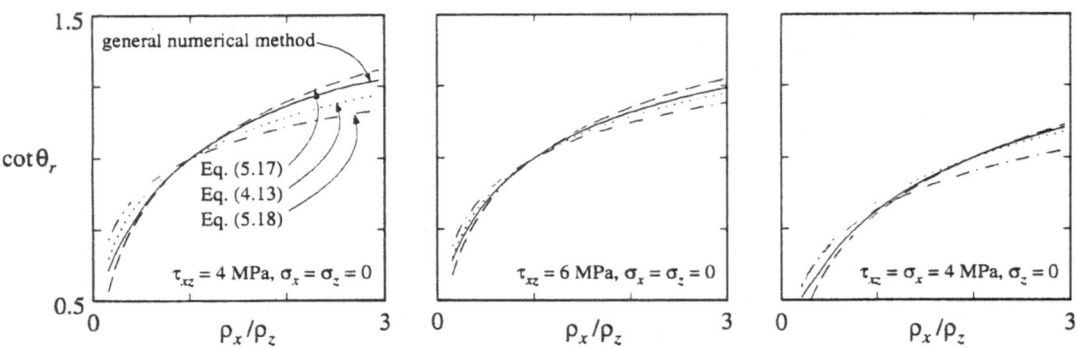

Fig. 5.5 – Cracked membrane model: crack directions according to different solutions.

be attributed to tension splitting, see Chapter 5.2.5. After the peak load, the reinforcement unloads elastically, with an initially steeper slope due to inverted bond action (for rigid-perfectly plastic bond behaviour). A secant stiffness value of the actual unloading curve [93,6] has been used in the calculations. The crack spacings increase slightly during the loading process; note that fictitious, rotating cracks are considered. The crack spacings are approximately equal to $s_{rmz} \approx s_{rmzo}/4$ and $s_{rmx} \approx 2s_{rmxo}$ and hence, tension stiffening has a pronounced effect on the x-reinforcement, while there is only little influence on the z-reinforcement. The crack widths w_r indicated in Fig. 5.4 are commented on in Chapter 5.2.5.

5.2.4 Approximate Analytical Solution

Assuming that the stresses and strains at the quarter points between the cracks, see Fig. 5.1 (f), are representative for the behaviour of the element, an approximate analytical solution can be established. Considering Figs. 5.1 (d) and (e) as well as Eqs. (5.2), (5.6) and (2.24), the steel stresses at the quarter points between the cracks follow from

$$\rho_x \sigma_{sx} = \sigma_x + \tau_{xz}\cot\theta_r - \frac{f_{ct}}{2}\lambda_x(1-\rho_x) \qquad \rho_z \sigma_{sz} = \sigma_z + \tau_{xz}\tan\theta_r - \frac{f_{ct}}{2}\lambda_z(1-\rho_z) \qquad (5.13)$$

Using the same approximation as for Eq. (5.10), Figs. 5.3 (b) and (c), the principal concrete compressive stress at the centre between the cracks can be expressed as

$$-\tau_{xz}(\tan\theta_r + \cot\theta_r) + f_{ct}(\lambda_x + \lambda_z)/2 + f_{ct}(\lambda_x - \lambda_z)\cos(2\theta_r)/2 \qquad (5.14)$$

Hence, like σ_{c1}, the principal compressive stresses in the concrete vary linearly between the cracks, see the comments regarding Eq. (5.10), page 70, and thus, the variation of $\cot\theta$ between the cracks is also linear. The principal compressive stresses in the concrete at the quarter points between the cracks are equal to

$$\sigma_{c3} = -\tau_{xz}(\tan\theta_r + \cot\theta_r) + \frac{f_{ct}}{2}(\lambda_x + \lambda_z - \lambda) \qquad (5.15)$$

where Eq. (5.11) has been used.

Assuming a linear elastic response the strains ε_x, ε_z, and ε_3 are determined by

$$\varepsilon_x = \frac{\sigma_x + \tau_{xz}\cot\theta_r - f_{ct}\lambda_x(1-\rho_x)/2}{\rho_x E_s} \qquad \varepsilon_z = \frac{\sigma_z + \tau_{xz}\tan\theta_r - f_{ct}\lambda_z(1-\rho_z)/2}{\rho_z E_s}$$

$$\varepsilon_3 = -\frac{\tau_{xz}(\tan\theta_r + \cot\theta_r) - f_{ct}(\lambda_x + \lambda_z - \lambda)/2}{E_c} \tag{5.16}$$

if the influence of lateral tensile stresses σ_{c1} in the concrete on ε_3 (Poisson's ratio) is neglected. Substituting the expressions (5.16) into the strain-compatibility condition $\cot^2\theta_r = (\varepsilon_z - \varepsilon_3)/(\varepsilon_x - \varepsilon_3)$, one obtains the equation

$$\tan^2\theta_r \rho_x (1 + n\rho_z) + \tan\theta_r \rho_x \left\{ \frac{\sigma_z}{\tau_{xz}} - \frac{f_{ct}}{2\tau_{xz}} \left[\lambda_z + n\rho_z \left(\lambda_x + \frac{n-1}{n}\lambda_z - \lambda \right) \right] \right\} = \tag{5.17}$$

$$\cot^2\theta_r \rho_z (1 + n\rho_x) + \cot\theta_r \rho_z \left\{ \frac{\sigma_x}{\tau_{xz}} - \frac{f_{ct}}{2\tau_{xz}} \left[\lambda_x + n\rho_x \left(\lambda_z + \frac{n-1}{n}\lambda_x - \lambda \right) \right] \right\}$$

for the crack inclination θ_r. Note that in general, λ_x and λ_z depend on θ_r according to Eqs. (5.5) and (5.12), and hence, Eq. (5.17) has to be solved in an iterative numerical manner.

For $f_{ct} = 0$, Eq. (5.17) reduces to Eq. (4.13). On the other hand, similar to Vecchio and Collins' [156] modified compression field approach, one may neglect the variation of the principal direction of the concrete stresses between the cracks and consider tensile stresses acting perpendicularly to the cracks, varying linearly from zero at the cracks to λf_{ct} at the centre between the cracks, where $0.5 \leq \lambda \leq 1$ according to Eq. (5.9). Thus, $\lambda_x = \lambda_z = \lambda$ and hence

$$\tan^2\theta_r \rho_x (1 + n\rho_z) + \tan\theta_r \rho_x \left\{ \frac{\sigma_z}{\tau_{xz}} - \frac{f_{ct}}{2\tau_{xz}}\lambda[1 + (n-1)\rho_z] \right\} = \tag{5.18}$$

$$\cot^2\theta_r \rho_z (1 + n\rho_x) + \cot\theta_r \rho_z \left\{ \frac{\sigma_x}{\tau_{xz}} - \frac{f_{ct}}{2\tau_{xz}}\lambda[1 + (n-1)\rho_x] \right\}$$

In Fig. 5.5 the solutions according to the general numerical method and Eqs. (5.17), (4.13) and (5.18) are compared for three particular stress combinations, assuming $\lambda = 1$, $\rho_z = 2\%$, $\varnothing_x = \varnothing_z = 16$ mm, $n = 10$, $f_c' = 30$ MPa and $\tau_{bo}/2 = f_{ct} = 0.3(f_c')^{2/3}$ in MPa. Note that there is no yielding of the reinforcement since linear elastic behaviour has been assumed, Eq. (5.16). For the calculations according to the general numerical method Eqs. (5.3) and (5.4) have been applied along with $\varepsilon_{co} = 0.002$.

From Fig. 5.5 it can be seen that Eq. (5.17) and even Eq. (4.13) closely approximate the solutions obtained from the general numerical method. On the other hand, starting from Eq. (4.13) (unbonded reinforcement), Eq. (5.18) (tension stiffening accounted for similarly as in [156,57]) changes $\cot\theta_r$ away from, rather than towards the solutions according to the general numerical method.

5.2.5 Crack Widths and Effect of Poisson's Ratio

The crack widths w_r can be determined from the average tensile strain due to cracks $\varepsilon_1^{(r)}$ and the diagonal crack spacing s_{rm} as $w_r = s_{rm}\varepsilon_1^{(r)}$, see Chapter 4.3.1. Noting that the principal directions of the average total strains $\boldsymbol{\varepsilon}$, of the strains due to cracks $\boldsymbol{\varepsilon}^{(r)}$, and of the concrete strains $\boldsymbol{\varepsilon}^{(c)}$ coincide and that $\varepsilon_1 = \varepsilon_x + \varepsilon_z - \varepsilon_3$, one obtains

$$\varepsilon_1^{(r)} = \frac{w_r}{s_{rm}} = \varepsilon_1 - \varepsilon_1^{(c)} = \varepsilon_x + \varepsilon_z - \varepsilon_3 - \varepsilon_1^{(c)} \tag{5.19}$$

see Eq. (4.10). The total compressive strain ε_3 is equal to the concrete compressive strain, $\varepsilon_3 = \varepsilon_3^{(c)}$, since $\varepsilon_3^{(r)} = 0$. Assuming a linear elastic behaviour and accounting for the influence of lateral stresses, the concrete strains $\boldsymbol{\varepsilon}^{(c)}$ can be expressed as

$$\varepsilon_1^{(c)} = (\sigma_{c1} - \nu_{13}\sigma_{c3})/E_c \qquad \varepsilon_3^{(c)} = (\sigma_{c3} - \nu_{31}\sigma_{c1})/E_c \tag{5.20}$$

where generally, the Poisson's ratios ν_{13} and ν_{31} have different values. The rotation of the principal direction of the concrete stresses and the variation of σ_{c1} and σ_{c3} between the cracks have been neglected in Eq. (5.20). As for the approximate analytical solution, Chapter 5.2.4, σ_{c1} and σ_{c3} can be taken as the stresses and strains at the quarter points between the cracks; this approximation is appropriate since the influence of Poisson's ratio is small as demonstrated below. For moderate concrete compressive stresses σ_{c3}, the concrete tensile strain $\varepsilon_1^{(c)}$ is fairly well predicted by Eq. $(5.20)_1$, using a Poisson's ratio of $\nu_{13} \approx 0.15$. The value of $\varepsilon_1^{(c)}$ does not affect the calculations according to the cracked membrane model since steel and concrete stresses depend on the total strains $\boldsymbol{\varepsilon}$, while the subdivision of ε_1 into $\varepsilon_1^{(c)}$ and $\varepsilon_1^{(r)}$ is not needed, Chapter 4.3.1. For crack width calculations according to Eq. (5.19), the value of $\varepsilon_1^{(c)}$ can thus be determined from Eq. (5.20) after the calculations for any given set of applied stresses. On the other hand, the influence of the lateral tensile stresses σ_{c1} in the concrete on $\varepsilon_3 = \varepsilon_3^{(c)}$ should basically be included in the calculations according to the cracked membrane model. However, this effect is far from linear elastic as presumed by Eq. (5.20). Rather, the deformed reinforcing bars tend to expand the surrounding concrete (see Chapter 2.4.2, tension splitting); lateral tensile strains rather than stresses might be the governing parameter of this phenomenon. If it is modelled by Eq. $(5.20)_2$, negative values of ν_{31} are obtained, even for comparatively low stresses in the reinforcement; lateral expansions due to tension splitting of roughly $\Delta\varepsilon_3 \approx 0.2...0.5‰$ (corresponding to "Poisson's ratios" of $\nu_{31} \approx -0.06$) have been observed in orthogonally reinforced panels at onset of yielding of the reinforcement [159]. Tension splitting thus results in a reduction of $|\varepsilon_3|$ by $\Delta\varepsilon_3$. This observation is supported by the fact that values of $|\varepsilon_3|$ at the peak compressive stress observed in orthogonally reinforced panels are typically lower than ε_{co} in uniaxial compression [155], see also Chapter 5.2.3, and could be accounted for by adopting a suitably reduced value of ε_{co}. However, since $\Delta\varepsilon_3$ is small compared to ε_x, ε_z, and ε_1, it has little impact on the overall response and is neglected in the cracked membrane model. In the approximate analytical solution presented below, ν_{31} will be included according to Eq. (5.20) in order to demonstrate its minor influence.

Having determined the total strains as well as the diagonal crack spacing s_{rm} from the cracked membrane model, $\varepsilon_1^{(r)}$ and w_r follow from Eq. (5.19), where $\varepsilon_1^{(c)}$ is determined by Eq. (5.20)$_1$.

An approximate analytical solution for the crack width in the linear elastic range can also be obtained. The stresses at the quarter points between the cracks follow from Eqs. (5.13) and (5.14), and $\sigma_{c1} = \lambda f_{ct}/2$. For linear elastic behaviour, the total strains ε_x and ε_z at the quarter points between the cracks follow from Eqs. (5.16)$_{1,2}$, while the concrete strains at the quarter points between the cracks are equal to

$$\varepsilon_3 = -\frac{\tau_{xz}(\tan\theta_r + \cot\theta_r) - f_{ct}[\lambda_x + \lambda_z - \lambda(1 + \nu_{31})]/2}{E_c}$$

$$\varepsilon_1^{(c)} = \frac{\lambda f_{ct}}{2E_c} + \nu_{13}\frac{\tau_{xz}(\tan\theta_r + \cot\theta_r) - f_{ct}(\lambda_x + \lambda_z - \lambda)/2}{E_c} \tag{5.21}$$

if Poisson's effect is taken into account according to Eq. (5.20). Thus, the tensile strains due to cracks $\varepsilon_1^{(r)}$ and hence, the crack width w_r follow from Eq. (5.19) as

$$\frac{w_r}{s_{rm}} = \frac{\sigma_x + (\tau_{xz}\cot\theta_r - \lambda_x f_{ct}/2)[1 + n\rho_x(1 - \nu_{13})]}{\rho_x E_s} + \tag{5.22}$$

$$\frac{\sigma_z + (\tau_{xz}\tan\theta_r - \lambda_z f_{ct}/2)[1 + n\rho_z(1 - \nu_{13})]}{\rho_z E_s} + \frac{f_{ct}}{2E_s}[\lambda_x + \lambda_z - n\lambda(\nu_{13} - \nu_{31})]$$

Eq. (5.22) reveals that the crack width is primarily controlled by the average strains of the reinforcements, ε_x and ε_z. Tension stiffening effects, represented by the expressions $-\lambda_x f_{ct}/2$ and $-\lambda_z f_{ct}/2$ in the numerators of the first two terms on the right-hand side of Eq. (5.22), typically result in reductions of w_r by about 20%. The concrete compressive strains contribute relatively little to w_r, particularly for small reinforcement ratios $\rho \ll 1$. The influence of the concrete tensile strain $\varepsilon_1^{(c)}$, including Poisson's effect, is negligible for $\rho \ll 1$ as confirmed by the crack widths calculated in the numerical example illustrated in Fig. 5.4; the crack widths including and excluding $\varepsilon_1^{(c)}$ are almost identical. For small reinforcement ratios, the crack widths can thus be calculated from the total tensile strain and the diagonal crack spacing, $w_r = s_{rm}\varepsilon_1$.

If the influence of the lateral tensile stresses σ_{c1} on the principal compressive strain $\varepsilon_3 = \varepsilon_3^{(c)}$ is accounted for according to Eq. (5.20), one obtains the expression

$$\tan^2\theta_r \rho_x (1 + n\rho_z) + \tan\theta_r \rho_x \left\{ \frac{\sigma_z}{\tau_{xz}} - \frac{f_{ct}}{2\tau_{xz}}\left[\lambda_z + n\rho_z\left(\lambda_x + \frac{n-1}{n}\lambda_z - \lambda(1 + \nu_{31}) \right) \right] \right\} = \tag{5.23}$$

$$\cot^2\theta_r \rho_z (1 + n\rho_x) + \cot\theta_r \rho_z \left\{ \frac{\sigma_x}{\tau_{xz}} - \frac{f_{ct}}{2\tau_{xz}}\left[\lambda_x + n\rho_x\left(\lambda_z + \frac{n-1}{n}\lambda_x - \lambda(1 + \nu_{31}) \right) \right] \right\}$$

A comparison with Eq. (5.17) reveals the marginal influence of $\nu_{31} \approx -0.06$, particularly for small reinforcement ratios.

5.3 Comparison with Previous Work

5.3.1 Comparison with Compression Field Approaches

Similar to the modified compression field approach proposed by Vecchio and Collins [156] and Hsu's [57] so-called rotating angle softened truss model, the cracked membrane model considers fictitious, rotating and stress-free cracks with a finite spacing and expresses compatibility conditions in terms of the average total strains. Compression softening effects are dealt with in essentially the same manner, see also Chapter 2.4.3, but the treatment of tension stiffening effects is entirely different.

In the cracked membrane model, crack spacings and tensile stresses between the cracks are determined from equilibrium conditions and bond shear stress-slip relationships. This allows accounting for different tension stiffening properties in the reinforcement directions. According to the cracked membrane model, tension stiffening has a more pronounced effect on the stronger reinforcement as demonstrated by the numerical example illustrated in Fig. 5.4. This is due to the fact that the cracks are generally oriented closer to the direction of the stronger reinforcement, while the crack spacings are connected by the geometrical condition (5.5). Hence, the crack spacing in the direction of the stronger reinforcement is normally larger than in the direction of the weaker reinforcement. For the crack spacings in uniaxial tension in the reinforcement directions, the opposite is normally the case, see Eq. (5.7). Thus, the stronger reinforcement is stiffened much more than the weaker reinforcement, resulting in crack inclinations closer to the direction of the stronger reinforcement than if tension stiffening had the same effect on both reinforcements, as implied by previous modified compression field approaches [156,57], see Fig. 5.5. Indeed, the principal compressive strain directions observed in tests are typically closer to the direction of the stronger reinforcement [67] than predicted by the modified compression field approach proposed by Vecchio and Collins [156].

The cracked membrane model eliminates the need for empirical constitutive equations relating average steel and concrete stresses and average strains in tension since it considers stresses at the crack rather than average stresses between the cracks. This allows re-introducing the link to limit analysis that had been lost with Vecchio and Collins' [156] modified compression field approach. In approaches that consider average stresses between the cracks [156,57], the equilibrium conditions can be expressed as

$$\sigma_x = \rho_x \sigma_{sxm} + \sigma_{c1m} \sin^2\theta + \sigma_{c3m} \cos^2\theta$$

$$\sigma_z = \rho_z \sigma_{szm} + \sigma_{c1m} \cos^2\theta + \sigma_{c3m} \sin^2\theta \qquad (5.24)$$

$$\tau_{xz} = (\sigma_{c1m} - \sigma_{c3m}) \sin\theta \cos\theta$$

where σ_{sxm}, σ_{szm}, σ_{c1m} and σ_{c3m} are the average stresses in the reinforcement and in the concrete between the cracks. The principal direction of the average concrete stresses between the cracks and the average total strains coincide and are both equal to θ. Rearranging Eq. (5.24) similar to Eq. (5.2) one gets the following expressions in terms of average steel and concrete stresses between the cracks

$$\sigma_{sxm} = (\sigma_x + \tau_{xz}\cot\theta - \sigma_{c1m})/\rho_x$$

$$\sigma_{szm} = (\sigma_z + \tau_{xz}\tan\theta - \sigma_{c1m})/\rho_z \qquad (5.25)$$

$$\sigma_{c3m} = \sigma_{c1m} - \tau_{xz}(\cot\theta + \tan\theta)$$

A comparison with Eq. (5.2) reveals that apart from being expressed in terms of average stresses between the cracks, Eq. (5.25) incorporates the tensile stresses in the concrete as an additional unknown; this has some important consequences.

In the original compression field approaches and in the cracked membrane model, ultimate loads can be determined similarly as in limit analysis, see also Chapter 5.3.2. In these models, calculation of the applied shear stress τ_{xz} at ultimate for given values of the applied axial stresses σ_x and σ_z involves five unknowns, i.e., τ_{xz}, σ_{sxr}, σ_{szr}, σ_{c3r} and θ_r, see Eq. (5.2). At the ultimate state, two of these unknowns are typically known (e.g. both reinforcements are yielding), and hence, the remaining three unknowns can be determined from the equilibrium conditions, Eq. (5.2). Due to the additional unknown σ_{c1m} in Eq. (5.25), this is impossible in approaches considering average stresses between the cracks [156,57] and hence, the link to limit analysis has been lost. In these models, ultimate loads cannot be determined without performing a complete load-deformation analysis, even in the simplest case of failure governed by yielding of both reinforcements.

The additional unknown σ_{c1m} also affects the interpretation of experimental results. If the applied stresses σ_x, σ_z and τ_{xz} as well as the average total strains are measured, the remaining unknown stresses at the cracks according to the cracked membrane model, σ_{sxr}, σ_{szr}, and σ_{c3r}, follow from Eq. (5.2). Within the assumptions of the cracked membrane model, the state of stress is thus completely determined by the equilibrium conditions. On the other hand, due to the additional unknown σ_{c1m}, average stresses cannot be determined from Eq. (5.25) alone; further assumptions have to be made. Empirical constitutive relationships between average steel stresses and average strains could be postulated for this purpose; from Eq. $(5.25)_{1,2}$ it can be seen that generally, the assumption of such relationships in both reinforcement directions would result in inconsistencies. Therefore, constitutive equations relating average concrete tensile stresses σ_{c1m} and average strains are typically assumed [156,57], see Chapter 2.4.2. However, relationships involving average tensile stresses in the concrete or in the reinforcement depend on the tensile strength of concrete, on the crack spacings and bond shear stresses as well as on the type of reinforcement used and are subject to rather wide scatter.

The deviation of the proposed compression softening relationship, Eq. (5.4), from the relationships introduced by Vecchio and Collins [156], Eq. (2.25), and by Hsu [57] is motivated by two observations. First, the cracked membrane model considers maximum stresses σ_{c3r} at the cracks, whereas previous modified compression field approaches consider average stresses σ_{c1m} and σ_{c3m} in the concrete between the cracks. Thus, according

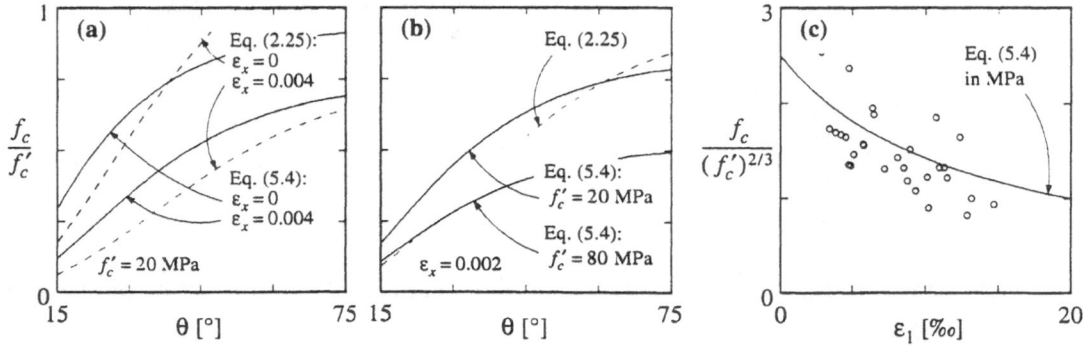

Fig. 5.6 – Compression softening relationships: (a), (b) comparison of proposed relationship, Eq. (5.4), with Eq. (2.25) suggested by Collins et al. [156,33]; (c) comparison of Eq. (5.4) with experimental evidence. Note: f'_c in MPa.

to the third Eq. (5.2) and the third Eq. (5.25), concrete compressive stresses according to the cracked membrane model are higher by σ_{c1m} if the same tests are interpreted (supposing that the angles θ_r or θ are known from strain measurements). Also, some tests used by Vecchio and Collins [156] in calibrating their compression softening relationship are predicted to fail by yielding of both reinforcements according to the cracked membrane model; they have been excluded from the calibration of Eq. (5.4), see Appendix B. Second, experimental evidence (Chapter 2.4.3) suggests a more drastic softening for high-strength concrete and hence, f_c is assumed to be proportional to $(f'_c)^{2/3}$ rather than f'_c. Along with these changes and compared to Eq. (2.25), the terms in the denominator on the right hand side of Eq. (5.4) have been somewhat rearranged. Figs. 5.6 (a) and (b) compare the proposed compression softening relationship with the expression suggested by Collins et al. [156,33], assuming $\varepsilon_3 = -\varepsilon_{co} = -0.002$; the principal tensile strains ε_1 have been calculated from $\varepsilon_1 = \varepsilon_x + (\varepsilon_x - \varepsilon_3)\cot^2\theta$. Fig. 5.6 (c) shows a comparison of Eq. (5.4) with experimental evidence; underlying test data and calculations are given in Appendix B. The agreement is satisfactory; note that only tests in which one or both reinforcements remained elastic at the ultimate state have been considered and that the limitation $f_c \leq f'_c$ in Eq. (5.4) has been suppressed in Fig. 5.6 (c). As further outlined in Appendix B, the principal tensile strain ε_1 and the concrete strength f'_c affect f_c to a comparable extent and hence, Eq. (5.4) agrees better with the experimental data obtained from panels of different concrete strengths than Eq. (2.25), which is independent of f'_c.

The approximate analytical solution derived in Chapter 5.2.4, Eq. (5.17), comprises the solution derived by Baumann [10], Eq. (4.13), as a special case. As another special case, Eq. (5.17) contains Eq. (5.18), which – for sufficiently small stresses – very closely approximates the crack inclination according to Vecchio and Collins' [156] modified compression field approach as well as Hsu's [57] so-called rotating angle softened truss model. Finally, the derivation underlying Eq. (5.17) also results in expression (5.12) for the diagonal crack spacing, which was suggested by Vecchio and Collins [156] without further justification.

5.3.2 Relation to Limit Analysis

The determination of failure conditions is often the most important aspect in design practice. Comprehensive load-deformation analyses as obtained from the cracked membrane model are inefficient in such situations. Assuming perfectly plastic behaviour of the concrete and the reinforcement, the determination of failure conditions according to limit analysis methods is straightforward, see Chapter 4.2. Yet as pointed out in Chapter 3.3.2, it is not easy to assess the effective concrete compressive strength f_c to be used in such calculations. In this chapter, approximations of f_c in terms of the effective reinforcement ratios in shear, $\tilde{\rho}_x = \rho_x - \sigma_x / f_{syx}$ and $\tilde{\rho}_z = \rho_z - \sigma_z / f_{syz}$, will be presented.

Figs. 5.7 (a) and (b) illustrate failure surfaces calculated according to limit analysis and the cracked membrane model, assuming identical steel properties in both reinforcement directions, $f_{sy} = 500$ MPa, $f_{su} = 625$ MPa, $E_s = 200$ GPa, $\varepsilon_{su} = 0.05$, $\lambda = 1$, $\varnothing = 16$ mm, $f_c' = 30$ MPa, $\varepsilon_{co} = 0.002$ and $\tau_{b1} = \tau_{bo}/2 = f_{ct} = 0.3(f_c')^{2/3}$ in MPa. The failure surface according to limit analysis, Fig. 5.7 (a), has been calculated assuming a constant value of the effective concrete compressive strength f_c. Since τ_{xz} is plotted versus $\tilde{\rho}_x$ and $\tilde{\rho}_z$ rather than against σ_x and σ_z as in Fig. 4.4 (d), Regimes 5 to 7 are not involved as long as $-\sigma_x \leq f_c + \rho_x f_{syx}' - \tilde{\rho}_z f_{syz}$ and $-\sigma_z \leq f_c + \rho_z f_{syz}' - \tilde{\rho}_x f_{syx}$. In Regime 1 both reinforcements yield, in Regime 4 the concrete crushes while both reinforcements remain elastic, and in Regimes 2 and 3, the weaker reinforcement yields and the concrete crushes. Ultimate loads τ_{xz} and the inclination θ of the principal compressive direction with respect to the x-axis are determined by Eqs. (4.4) and (4.7).

Fig. 5.7 (b) shows the failure surface obtained from the general numerical method of the cracked membrane model. The dash-dotted lines correspond to the combinations of the effective reinforcement ratios $\tilde{\rho}_x$ and $\tilde{\rho}_z$ below which failure is governed by rupture of the weaker reinforcement. For values of $\tilde{\rho}_x$ and $\tilde{\rho}_z$ above these lines, failure occurs by crushing of the concrete, either before or after yielding of the reinforcement(s). The dashed lines indicate the combinations of the effective reinforcement ratios $\tilde{\rho}_x$ and $\tilde{\rho}_z$ for which at the ultimate state, the maximum steel stress at the cracks is equal to the yield stress. Hence, inside the region enclosed by the dashed lines, the maximum steel stresses in both reinforcements are above the yield stress at the ultimate state. Failures within this region can be attributed to yielding of both reinforcements although, strictly speaking, collapse eventually occurs by crushing of the concrete or by rupture of the weaker reinforcement while both reinforcements are yielding. However, ultimate loads in this region are only slightly higher than predicted by limit analysis for yielding of both reinforcements, Eq. $(4.4)_1$, see also Appendix B.

The results obtained from limit analysis and the cracked membrane model are quite similar. Fig. 5.8 compares the failure surfaces shown in Fig. 5.7 for three different sections, i.e. $\tilde{\rho}_x f_{syx} / (f_c')^{2/3} = 0.4$, 0.85 and 2.5 (with f_c' in MPa), respectively; the curves corresponding to Fig. 5.7 (c) will be commented on later. A value of $f_c = 1.71 (f_c')^{2/3}$ in MPa has been chosen, such that $\tau_{xz} = f_c/2$ in Regime 4 according to limit analysis is equal to the ultimate shear stress obtained from the cracked membrane model at the in-

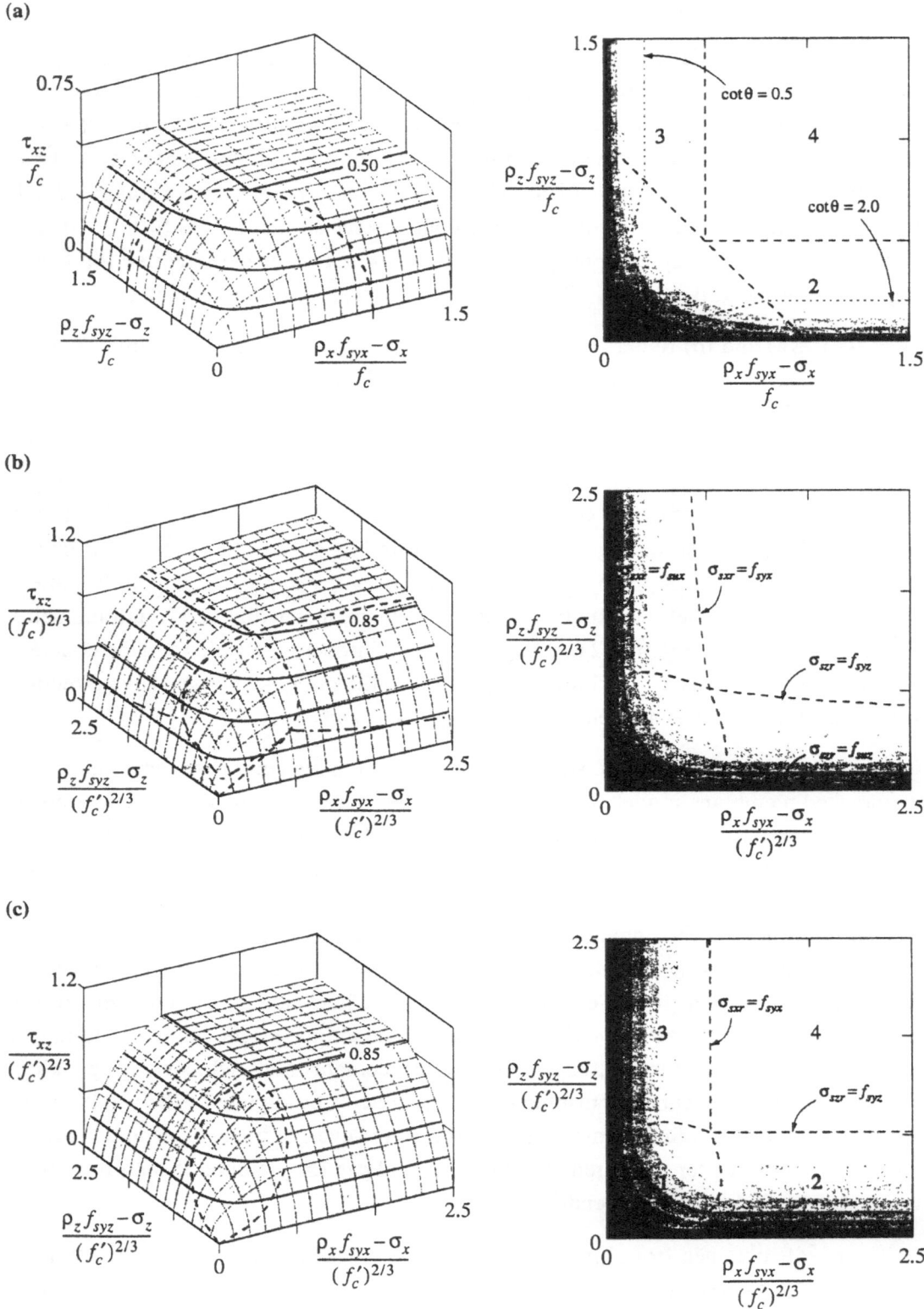

Fig. 5.7 – Failure surfaces: (a) limit analysis, f_c = constant; (b) cracked membrane model; (c) limit analysis, accounting for compression softening of the concrete, Eq. (5.33). Note: f'_c in MPa.

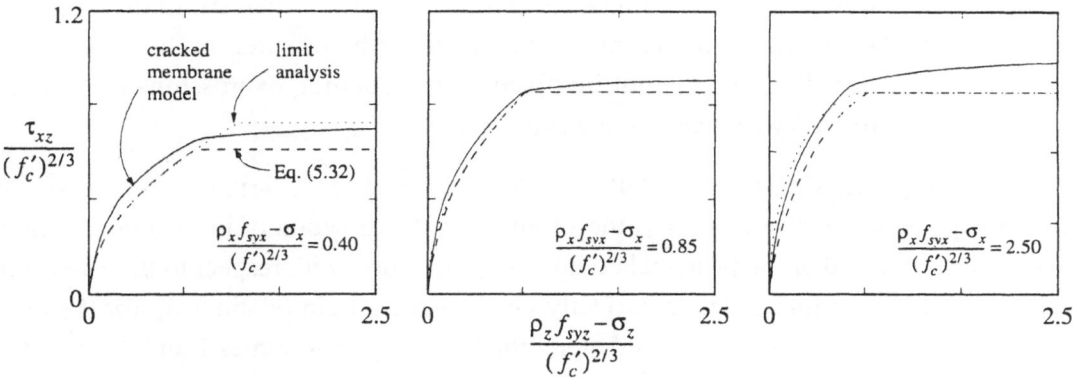

Fig. 5.8 – Comparison of failure surfaces shown in Fig. 5.7 at three different sections $(\rho_x f_{syx} - \sigma_x)/(f_c')^{2/3} = $ constant. Note: f_c' in MPa.

tersection of the dashed curves in Fig. 5.7 (b), i.e., for crushing of the concrete while $\sigma_{sxr} = f_{syx}$ and $\sigma_{szr} = f_{syz}$. From Fig. 5.8 it can be seen that limit analysis overestimates the ultimate load in the regions where compression softening of the concrete is most pronounced, i.e., for high values of the principal tensile strain ε_1 resulting from small or substantially different values of $\tilde{\rho}_x$ and $\tilde{\rho}_z$. According to the cracked membrane model, failures are governed by rupture of the weaker reinforcement only for very small values of $\tilde{\rho}$; if compression softening of the concrete is neglected, this type of failure becomes much more important [144]. In limit analysis, steel ruptures play no part because of the assumption of perfect plasticity. Design codes impose limitations on the inclination θ in order to avoid failures governed by steel rupture or excessive softening of the concrete; typically, a range of $\cot\theta$ between 0.5 and 2 is permitted, Fig. 5.7 (a). Apart from small values of $\tilde{\rho}$, the corresponding boundaries according to the cracked membrane model – the dashed and dash-dotted lines in Fig. 5.7 (b) – are not correctly represented by such limitations, and in Regimes 2 and 3, limit analysis still overestimates the ultimate load. On the other hand, failure loads obtained from the cracked membrane model exceed those obtained from limit analysis in many situations, primarily because of strain-hardening of the reinforcement. While estimates of the failure load can still be obtained from limit analysis by adopting a suitable value of f_c, the situation is not entirely satisfactory. Thürlimann et al. [152] therefore proposed to use $f_c = f_{co}\sin 2\theta$ in order to account for compression softening effects, where $f_{co} = f_c(\theta = \pi/4)$. This approach is not further considered here; rather, the expression for f_c used in the cracked membrane model, Eq. (5.4), will be adopted. Substituting $c_1 = 0.4$ and $c_2 = 30$, Eq. (5.4) transforms to

$$f_c = \frac{(f_c')^{2/3}}{c_1 + c_2\varepsilon_1} \leq f_c' \qquad \text{in MPa.} \tag{5.26}$$

According to this expression, f_c depends on the principal tensile strain ε_1 which cannot be determined from limit analysis. However, as outlined below, ε_1 can be expressed in terms of $\tilde{\rho}_x$ and $\tilde{\rho}_z$ if certain assumptions are made. Fig. 5.7 (c) shows a failure surface obtained in this way; the failure regimes are denoted as in limit analysis. In the considerations below, the stronger reinforcement is assumed to be oriented in the x-direction, $\tilde{\rho}_x > \tilde{\rho}_z$, and hence, Regime 3 is not involved; corresponding expressions can be obtained from Regime 2 by exchanging the subscripts x and z.

The principal tensile strain ε_1 follows from the condition $\varepsilon_1 = \varepsilon_x + (\varepsilon_x - \varepsilon_3)\cot^2\theta$ if the principal compressive strain ε_3, the strain ε_x in the direction of the x-reinforcement and the inclination θ of the principal compressive direction with respect to the x-axis are known. Assuming a linear elastic-perfectly plastic stress-strain relationship for the reinforcement, ε_1 can thus be determined along the boundary of Regimes 1 and 2 where the weaker reinforcement yields, the concrete crushes, and the x-reinforcement is at the onset of yielding. The inclination θ follows from Eq. (4.7)$_1$ since $\sigma_{sx} = f_{syx}$ and $\sigma_{sz} = f_{syz}$. Neglecting slight reductions of $|\varepsilon_3|$ compared to ε_{co}, Chapters 5.2.3 and 5.2.5, the principal compressive strain ε_3 is equal to $-\varepsilon_{co}$ since the concrete crushes. Furthermore, $\varepsilon_{sxr} = \varepsilon_{syx} = f_{syx}/E_s$ is known since the x-reinforcement is at the onset of yielding. Due to tension stiffening effects, the average strain ε_x is somewhat smaller than ε_{sxr}; typically, $\varepsilon_x \approx 0.8\varepsilon_{syx}$ at onset of yielding, see Chapter 2.4.2. Substituting the resulting expression for ε_1 into Eq. (5.26), one gets

$$f_c = \frac{(f_c')^{2/3}}{c_1 + c_2\left(\varepsilon_x + (\varepsilon_x - \varepsilon_3)\dfrac{\tilde{\rho}_x f_{syx}}{\tilde{\rho}_z f_{syz}}\right)} \tag{5.27}$$

at the boundary of Regimes 1 and 2, where $\varepsilon_3 = -\varepsilon_{co}$ and $\varepsilon_x \approx 0.8 f_{syx}/E_s$. The principal concrete compressive stress in Regime 1 is equal to $-\sigma_{c3} = \tilde{\rho}_x f_{syx} + \tilde{\rho}_z f_{syz}$, see Eqs. (4.4) and (4.7). Observing that it must not exceed f_c, one obtains

$$\frac{\tilde{\rho}_x f_{syx} + \tilde{\rho}_z f_{syz}}{(f_c')^{2/3}}\left[c_1 + c_2\left(\varepsilon_x + (\varepsilon_x - \varepsilon_3)\frac{\tilde{\rho}_x f_{syx}}{\tilde{\rho}_z f_{syz}}\right)\right] = 1 \tag{5.28}$$

for the boundary of Regimes 1 and 2. In Regimes 2 and 4, failure loads depend on f_c and thus on the principal tensile strain ε_1. For values of $\tilde{\rho}_x$ and $\tilde{\rho}_z$ higher than along the boundary to Regime 1, the steel stresses in the x-reinforcement (Regime 2) or in both reinforcements (Regime 4) are below the yield stress. The strains in the non-yielding reinforcements decrease for higher reinforcement ratios and hence, f_c increases. This effect is neglected below, assuming that the strains in the non-yielding reinforcements are equal to $\varepsilon \approx 0.8 f_{sy}/E_s$ throughout Regimes 2 and 4; this simplification is on the safe side since compression softening effects are overestimated. In Regime 4, the assumption that $\varepsilon \approx 0.8 f_{sy}/E_s$ results in $\cot\theta = 1$ and a constant value of f_c,

$$f_c = \frac{(f_c')^{2/3}}{c_1 + c_2(2\varepsilon_x - \varepsilon_3)} \tag{5.29}$$

where $\varepsilon_3 = -\varepsilon_{co}$ and $\varepsilon_x \approx 0.8 f_{syx}/E_s$ as outlined above. Along the boundary between Regimes 2 and 4, the (weaker) reinforcement in the z-direction is at the onset of yielding, i.e., $\sigma_{szr} = (\sigma_z + \tau_{xz}\tan\theta)/\rho_z = f_{syz}$ according to Eq. (5.2), and from Eq. (5.29) one obtains the condition

$$\frac{\tilde{\rho}_z f_{syz}}{(f_c')^{2/3}} = \frac{0.5}{c_1 + c_2(2\varepsilon_x - \varepsilon_3)} \tag{5.30}$$

In Regime 2, the concrete crushes, $\varepsilon_3 = -\varepsilon_{co}$ and $\varepsilon_x \approx 0.8\, f_{syx}/E_s$ according to the assumptions made above. The inclination θ follows from Eq. $(4.7)_2$ and thus, one obtains a quadratic equation for f_c, with the solution

$$f_c = \frac{\tilde{\rho}_z f_{syz}}{2c_2(\varepsilon_x - \varepsilon_3)}\left\{\sqrt{(c_1 + c_2\varepsilon_3)^2 + 4c_2(\varepsilon_x - \varepsilon_3)\frac{(f_c')^{2/3}}{\tilde{\rho}_z f_{syz}}} - (c_1 + c_2\varepsilon_3)\right\} \tag{5.31}$$

where $\varepsilon_x \approx 0.8\, f_{syx}/E_s$ and $\varepsilon_3 = -\varepsilon_{co}$.

Substituting the expressions for f_c, Eq. (5.27), Eq. (5.29) and Eq. (5.31), respectively, into the corresponding Eq. (4.4), one gets the following expressions for the ultimate loads in Regimes 1, 2 and 4 in terms of the effective reinforcement ratios in shear, $\tilde{\rho}_x = \rho_x - \sigma_x/f_{syx}$ and $\tilde{\rho}_z = \rho_z - \sigma_z/f_{syz}$:

$$Y_1: \tau_{xz}^2 = \tilde{\rho}_x f_{syx}\, \tilde{\rho}_z f_{syz}$$

$$Y_2: \tau_{xz}^2 = \frac{(\tilde{\rho}_z f_{syz})^2}{2c_2(\varepsilon_x - \varepsilon_3)}\left\{\sqrt{(c_1 + c_2\varepsilon_3)^2 + 4c_2(\varepsilon_x - \varepsilon_3)\frac{(f_c')^{2/3}}{\tilde{\rho}_z f_{syz}}} - [c_1 + c_2(2\varepsilon_x - \varepsilon_3)]\right\} \tag{5.32}$$

$$Y_4: \tau_{xz}^2 = \left(\frac{(f_c')^{2/3}}{c_1 + c_2(2\varepsilon_x - \varepsilon_3)}\right)^2 /4$$

where $\varepsilon_x \approx 0.8\, f_{syx}/E_s$ and $\varepsilon_3 = -\varepsilon_{co}$. Assuming typical values for the strains, $\varepsilon_x = 0.002$ and $\varepsilon_{co} = 0.002$, and substituting $c_1 = 0.4$, $c_2 = 30$, see Eq. (5.4), Eq. (5.32) reduces to

$$Y_1: \tau_{xz}^2 = \tilde{\rho}_x f_{syx}\, \tilde{\rho}_z f_{syz}$$

$$Y_2: \tau_{xz}^2 = (\tilde{\rho}_z f_{syz})^2\left(\sqrt{2.0 + \frac{25}{3}\frac{(f_c')^{2/3}}{\tilde{\rho}_z f_{syz}}} - \frac{29}{12}\right) \tag{5.33}$$

$$Y_4: \tau_{xz}^2 = \left\{\frac{25}{29}(f_c')^{2/3}\right\}^2$$

The failure surface plotted in Fig. 5.7 (c) is based on Eq. (5.33). Fig. 5.8 reveals that the failure loads determined from Eq. (5.33) are always on the safe side as compared to the cracked membrane model, even for very small effective reinforcement ratios, where failure according to the cracked membrane model is governed by rupture of the reinforcement. Differences are primarily due to strain-hardening of the reinforcing steel. The expressions (5.33) are quite simple and appear suitable for design purposes.

5.3.3 Correlation with Experimental Evidence

In this chapter, calculations according to the cracked membrane model are compared with experimental results obtained from tests on orthogonally reinforced concrete panels. The experiments were conducted in the University of Toronto's "Panel Tester" (Series PV, Vecchio and Collins [155]) and "Shell Element Tester" [68,70] (Series PP, Marti and Meyboom [91], and SE, Khalifa [68], Kirschner and Collins [70]), and in the University of Houston's "Universal Panel Tester" [58] (Series HB, Zhang [168], and VA/VB, Zhang and Hsu [169,170]). These testing facilities allow subjecting membrane elements to a reasonably homogeneous state of in-plane stresses, avoiding edge restraints as far as possible. The selected tests were chosen from all test series of which complete sets of strain measurements could be obtained. They cover a wide range of parameters such as reinforcement ratios, concrete strength, and partial prestressing. Where available, specimens in which at least one reinforcement remained elastic at the ultimate state have been selected. Generally, the response of panels that fail by yielding of both reinforcements can be much more accurately predicted than the behaviour of panels whose failure is governed by crushing of the concrete [30]. The diagrams presented in this chapter allow for a direct and realistic comparison of predicted and experimental responses; note that $\cot\theta_r$ of the experimental curves corresponds to the principal compressive direction of the average strains measured in the experiments.

In the calculations according to the cracked membrane model, the general numerical method has been rigorously applied as outlined in Chapter 5.2.3. In particular, bond shear stresses $\tau_{b1} = \tau_{bo}/2 = f_{ct} = 0.3(f_c')^{2/3}$ in MPa and the compression softening relationship (5.4) have been used in all calculations. The diagonal crack spacing has been de-

Panel	VA0	VA1	VA2	VA3	VA4	PV27	PV23	PV25	PV28	Units
Size	1397·1397·178 [a]					890·890·70				mm
$\sigma_x/|\tau_{xz}| = \sigma_z/|\tau_{xz}|$	0	0	0	0	0	0	-0.39	-0.69	+0.32	–
$\rho_x = \rho_z$	0.571 [c]	1.143 [c]	2.276 [c]	3.419 [c]	4.990 [c]	1.785	1.785	1.785	1.785	%
$\varnothing_x = \varnothing_z$	11.3	11.3	16.0	19.5	25.2	6.35	6.35	6.35	6.35	mm
$f_{syx} = f_{syz}$	445	445	409	455	470	442	518	466	483	MPa
$f_{sux} = f_{suz}$	579 [d]	579 [d]	534 [d]	608 [d]	606 [d]	508 [b]	596 [b]	536 [b]	555 [b]	MPa
$\varepsilon_{sux} = \varepsilon_{suz}$	100 [d]	100 [d]	100 [d]	100 [d]	100 [d]	100 [b]	100 [b]	100 [b]	100 [b]	‰
E_s	200	200	200	200	200	200	200	200	200	GPa
f_c'	98.8	95.1	98.2	94.6	103.1	20.5	20.5	19.3	19.0	MPa
ε_{co}	2.40	2.45	2.50	2.45	2.35	1.90	2.00	1.80	1.85	‰

[a] panel VA4: thickness 203 mm.   [b] assumption based on stress-strain diagram in test report.
[c] average over entire panel; values given in test report apply only to central portion of the panels and are 5% higher.
[d] f_{su} = stress measured at ε_s = 50‰; ε_{su} not given in test report, 100‰ = typical value for type of reinforcement used.

Tab. 5.1 – Basic data for calculations underlying Fig. 5.9.

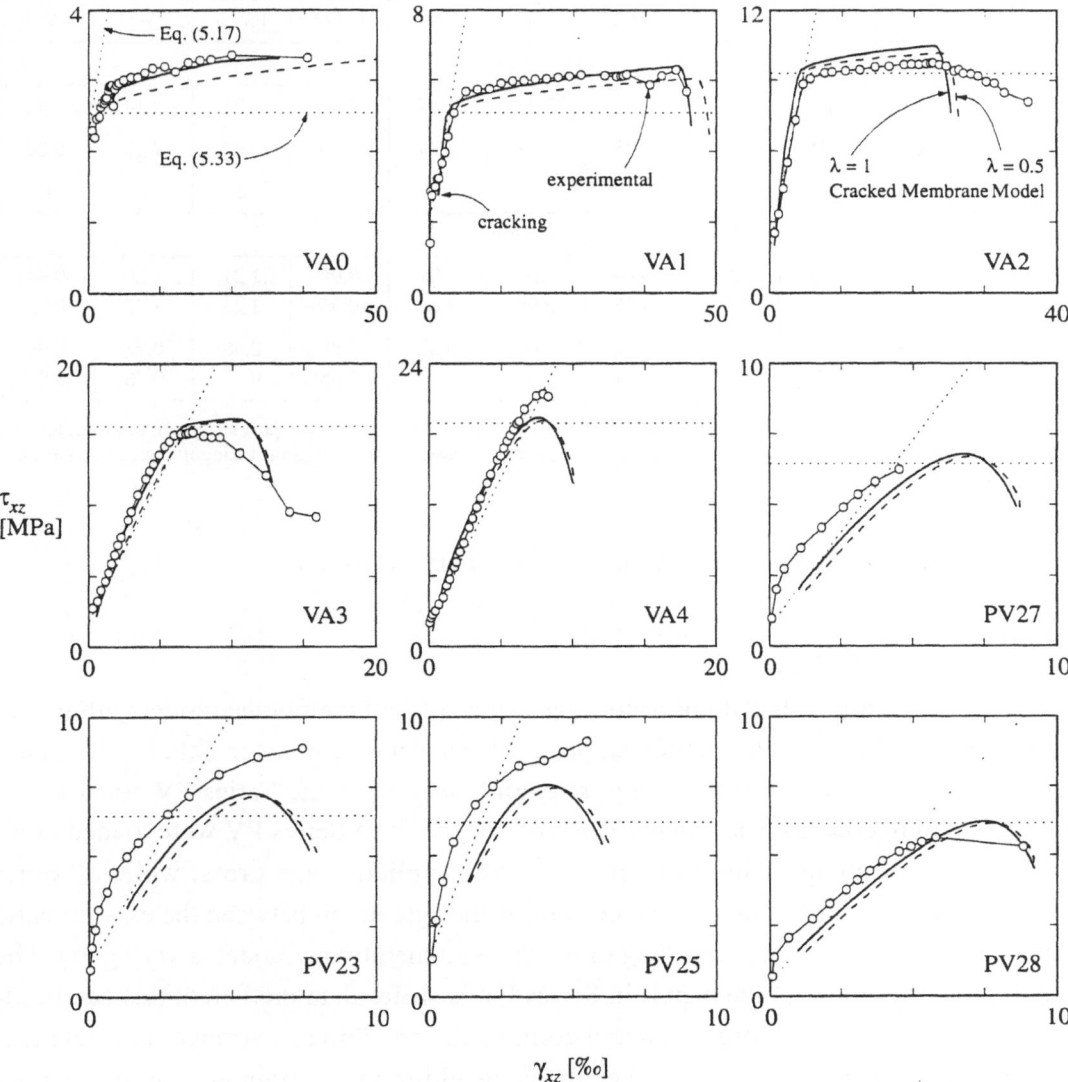

Fig. 5.9 – Comparison with experiments on isotropically reinforced concrete panels of Series PV by Vecchio and Collins [155] and Series VA by Zhang and Hsu [169,170].

termined from the general expression Eq. (5.8). The calculations have been performed for both $\lambda = 1$ (maximum diagonal crack spacing) and $\lambda = 0.5$ (minimum diagonal crack spacing). The treatment of unloading phenomena assumes elastic-plastic rather than hyperelastic properties, i.e., upon unloading, an elastic unloading rather than a return on the stress-strain curve takes place, see Fig. 5.4. For comparison purposes, ultimate loads determined from Eq. (5.33) are also given, along with the results of a cracked linear elastic analysis according to the approximate analytical solution presented in Chapter 5.2.4, Eq. (5.17), assuming a modular ratio of $n = 10$. Consideration of the initial, uncracked behaviour was excluded from the calculations.

Panel	VA0	VA1	VA2	VA3	VA4	PV27	PV23	PV25	PV28
Experiment [a]									
τ_r [MPa]	p	2.61	p	2.45	2.34	2.04	3.73	4.14	1.66
τ_{uexp} [MPa]	3.35	6.16	9.73	15.1	21.4	6.35	8.87	9.12	5.80
x- and z-reinf.	(u)	y	y	y	e	e	e	e	e
Calculation [b]									
τ_{uexp}/τ_{ucalc}	1.03	0.96	0.94	0.94	1.08	0.94	1.21	1.20	0.94
	1.10	1.02	0.96	0.95	1.10	0.95	1.23	1.22	0.95
$\sigma_{sxu}/f_{syx} = \sigma_{szu}/f_{syz}$	1.30 [c]	1.26	1.12	1.04	0.83	0.86	0.48	0.28	0.95
	1.30 [c]	1.19	1.09	1.02	0.82	0.85	0.47	0.28	0.94

[a] e / y: reinforcement elastic/yielding at concrete crushing; (u): steel rupture imminent (test stopped); p: pre-cracked.
[b] upper values: $\lambda = 1$ (maximum diagonal crack spacing); lower values: $\lambda = 0.5$ (minimum diagonal crack spacing).
[c] ($= f_{su}/f_{sy}$)

Tab. 5.2 – Isotropically reinforced panels: summary of results.

Fig. 5.9 compares calculations according to the cracked membrane model with the results of tests on isotropically reinforced panels; basic data are given in Tab. 5.1. The panels of Series VA were made from high-strength concrete, while Series PV consisted of normal-strength concrete specimens. Three of the panels in Series PV were loaded in biaxial tension or compression proportional to the applied shear stress, while all other specimens were loaded in pure shear. Generally, the agreement between the experimental results and the calculations according to the cracked membrane model is very good. The experimental response of the panels in Series PV is stiffer than predicted; this can be attributed to the fact that in order to avoid corner and edge failures, stronger concrete was cast around the perimeter of these specimens, resulting in a certain amount of restraint for the central portion of the panels and somewhat stiffer responses.

Tab. 5.2 compares the failure loads and the stresses in the reinforcement at the ultimate state according to the cracked membrane model with the experimentally observed ultimate loads and failure conditions. Biaxial compression in Panels PV23 and PV25 resulted in an increase of the ultimate load greater than predicted by the cracked membrane model. Apart from these two panels, the ultimate loads are predicted within 10%. Panels VA1, VA2 and VA3 all failed by yielding of the reinforcement and subsequent crushing of the concrete. Specimen VA0 could not be tested until failure because of the limited stroke of the actuators; according to the cracked membrane model, the panel would eventually have failed by rupture of the reinforcement. Specimen VA4 and all panels of Series PV failed by crushing of the concrete while the reinforcement was still elastic. From Tab. 5.2 it can be seen that failure modes are correctly predicted for all panels. The cracked linear elastic response according to Eq. (5.17) and the ultimate load obtained from Eq. (5.33) closely match the actual behaviour of all nine specimens.

Figs. 5.10 to 5.12 compare calculations according to the cracked membrane model with the results of tests on orthotropically reinforced panels. The stronger reinforcement was oriented in the x-direction in all specimens; basic data are given in Tab. 5.3. The panels of Series HB and SE were made from high-strength concrete, while all other specimens were made from normal-strength concrete. Specimens PP2 and PP3 were partially prestressed in the x-direction, and all specimens were loaded in pure shear. As for the isotropically reinforced panels, the agreement between the experimental results and the calculations according to the cracked membrane model is very good. Again, the response of the panels in Series PV was somewhat stiffer than in the other series, probably for the same reason as in the isotropically reinforced specimens.

Tab. 5.4 compares the failure loads and the stresses in the reinforcement at ultimate according to the cracked membrane model with the test results; the ultimate loads were predicted within 12% in all specimens. All panels of Series PV and SE as well as Specimens PP1 and VB3 failed by yielding of the z-reinforcement and crushing of the con-

Panel	PV10	PV19	PV20	PV21	PV22	PP1	PP2	PP3	SE1	SE6	HB3	HB4	VB1	VB2	VB3	Units
Size	890·890·70					1626·1626·287					1397·1397·178					mm
ρ_x	1.785	1.785	1.785	1.785	1.785	1.942	1.295	0.647	2.930	2.930	1.71[a]	2.84[a]	2.28[a]	3.42[a]	5.70[a]	%
$\varnothing_x$	6.35	6.35	6.35	6.35	6.35	19.5	16.0	11.3	19.5	19.5	19.5	25.2	16.0	19.5	25.2	mm
f_{syx}	276	458	460	458	458	479	486	480	492	492	446	470	409	455	470	MPa
f_{sux}	317[b]	527[b]	529[b]	527[b]	527[b]	667	630	640	640[b]	640[b]	583[c]	629[c]	534[c]	608[c]	606[c]	MPa
ε_{sux}	100[b]	100[b]	100[b]	100[b]	100[b]	90	100	91	100[b]	100[b]	100[c]	100[c]	100[c]	100[c]	100[c]	‰
ρ_z	0.999	0.713	0.885	1.296	1.524	0.647	0.647	0.647	0.978	0.326	0.57[a]	0.57[a]	1.14[a]	1.14[a]	1.14[a]	%
$\varnothing_z$	4.70	4.01	4.47	5.41	5.87	11.3	11.3	11.3	11.3	11.3	11.3	11.3	11.3	11.3	11.3	mm
f_{syz}	276	299	297	302	420	480	480	480	479	479	450	450	445	445	445	MPa
f_{suz}	317[b]	344[b]	341[b]	347[b]	483[b]	640	640	640	623[b]	623[b]	579[c]	579[c]	579[c]	579[c]	579[c]	MPa
ε_{suz}	100[b]	100[b]	100[b]	100[b]	100[b]	91	91	91	100[b]	100[b]	100[c]	100[c]	100[c]	100[c]	100[c]	‰
ρ_{px}	–	–	–	–	–	–	0.293	0.586	–	–	–	–	–	–	–	%
f_{pyx}	–	–	–	–	–	–	910	910	–	–	–	–	–	–	–	MPa
f_{pux}	–	–	–	–	–	–	1135	1135	–	–	–	–	–	–	–	MPa
ε_{pox}	–	–	–	–	–	–	3.535	3.750	–	–	–	–	–	–	–	‰
ε_{pux}	–	–	–	–	–	–	100	100	–	–	–	–	–	–	–	‰
$E_s = E_p$	200	200	200	200	200	200	200	200	200	200	200	200	200	200	200	GPa
f_c'	14.5	19.0	19.6	19.5	19.6	27.0	28.1	27.7	42.5	40.0	66.8	62.9	98.2	97.6	102.3	MPa
ε_{co}	2.70	2.15	1.80	1.80	2.00	2.12	2.38	1.92	2.54	2.50	2.80	2.70	2.50	2.45	2.35	‰

[a] average over entire panel; values given in test report apply only to central portion of the panels and are 5% higher.
[b] assumption based on stress-strain diagram in test report.
[c] f_{su} = stress measured at ε_s = 50‰; ε_{su} not given in test report, 100‰ = typical value for type of reinforcement used.

Tab. 5.3 – Basic data for calculations underlying Figs. 5.10 to 5.12.

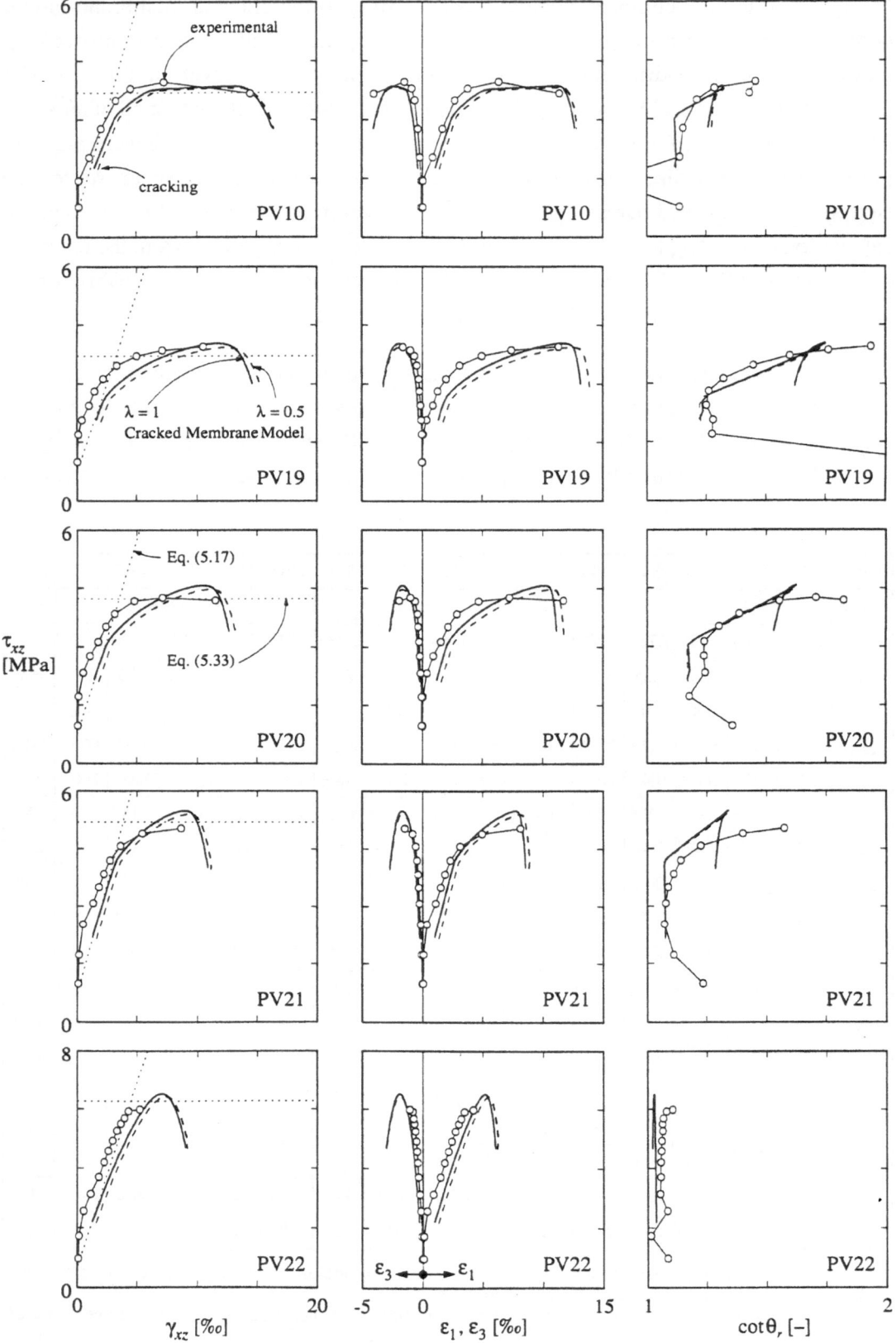

Fig. 5.10 – Comparison with experiments of Series PV by Vecchio and Collins [155].

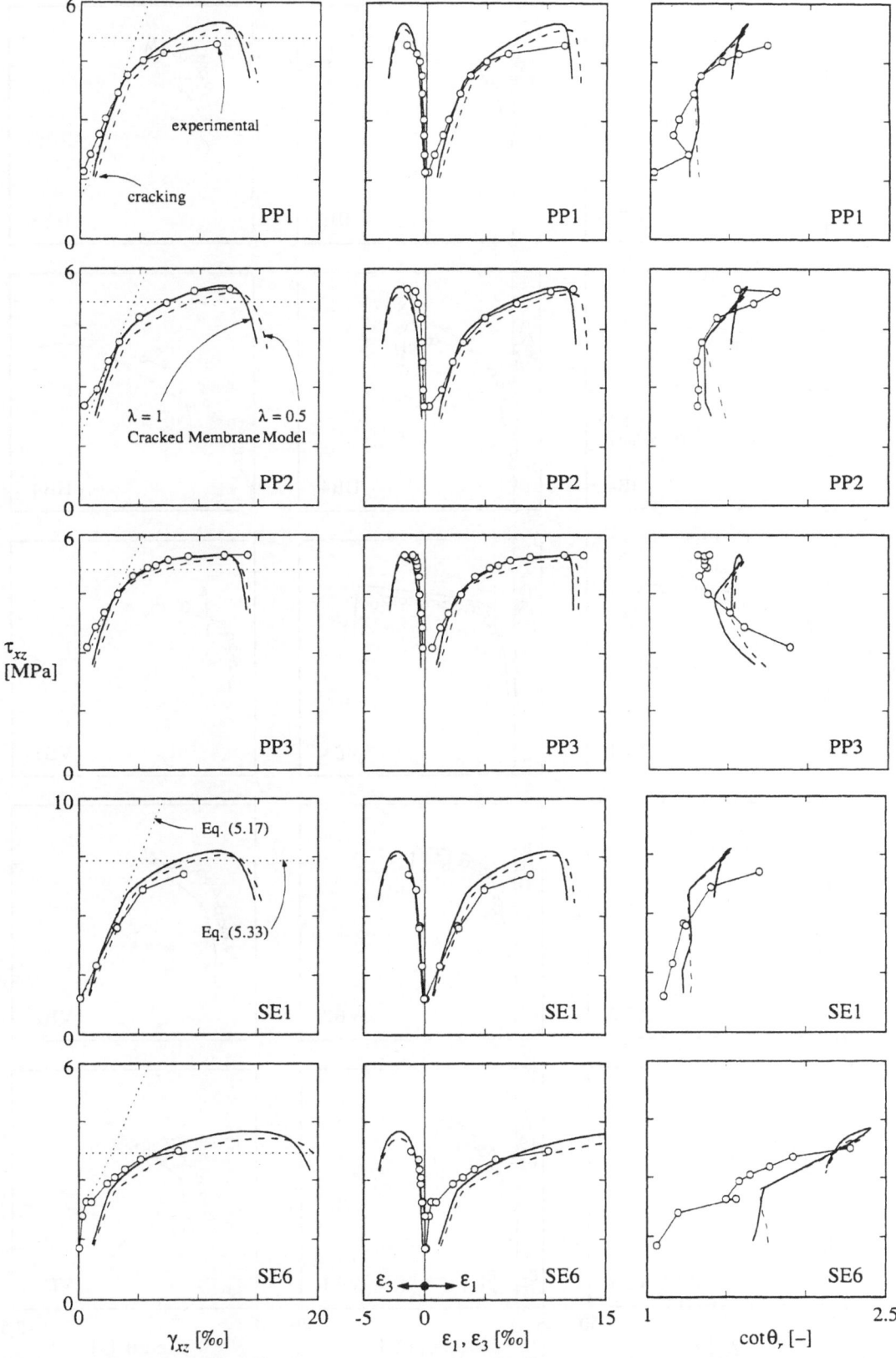

Fig. 5.11 – Comparison with experiments of Series PP by Marti and Meyboom [91] and Series SE by Khalifa [68], Kirschner and Collins [70].

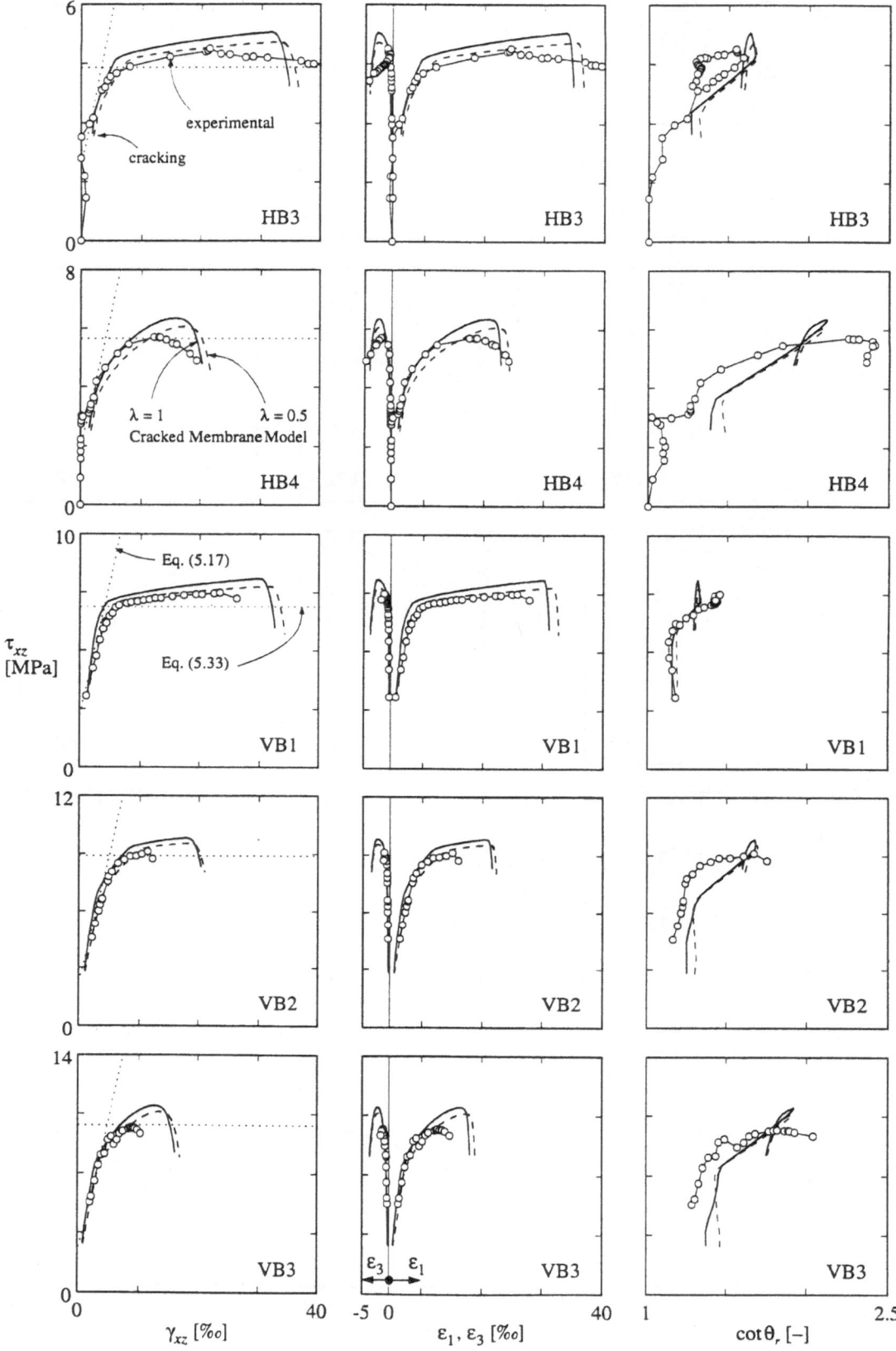

Fig. 5.12 – Comparison with experiments of Series HB by Zhang [168] and Series VB by Zhang and Hsu [169,170].

Panel	PV10	PV19	PV20	PV21	PV22	PP1	PP2	PP3	SE1	SE6	HB3	HB4	VB1	VB2	VB3
Experiment [a]															
τ_r [MPa]	1.86	2.07	2.21	2.35	2.42	1.71	2.54	3.15	n	n	2.65	3.02	3.09	2.47	2.57
τ_{uexp} [MPa]	3.97	3.95	4.26	5.03	6.07	4.95	5.50	5.50	6.76	3.75	4.89	5.71	7.50	9.14	9.71
x-reinf.	e	e	e	e	e	e	y	y	e	e	n	n	y	y	e
x-prestr.	–	–	–	–	–	–	y	y	–	–	–	–	–	–	–
z-reinf.	y	y	y	y	y	y	y	y	y	y	n	n	y	y	y
Calculation [b]															
τ_{uexp}/τ_{ucalc}	1.03	0.98	0.94	0.92	0.93	0.91	0.99	1.00	0.88	0.89	0.93	0.90	0.93	0.93	0.88
	1.04	1.00	0.95	0.93	0.94	0.93	1.02	1.02	0.89	0.93	0.97	0.94	0.97	0.96	0.92
σ_{sxu}/f_{syx}	1.03	0.86	0.90	0.90	0.82	0.95	1.01	1.06	0.82	0.71	1.14	1.00	1.14	1.06	0.79
	1.02	0.84	0.88	0.88	0.81	0.92	0.96	1.03	0.80	0.67	1.10	0.95	1.10	1.04	0.76
σ_{pxu}/f_{pyx}	–	–	–	–	–	–	1.01	1.01	–	–	–	–	–	–	–
							1.01	1.01							
σ_{szu}/f_{syz}	1.06	1.09	1.07	1.05	1.00	1.10	1.11	1.13	1.08	1.15	1.26	1.18	1.21	1.16	1.13
	1.04	1.06	1.05	1.03	0.98	1.08	1.08	1.09	1.06	1.11	1.19	1.13	1.15	1.11	1.09

[a] e / y: reinforcement elastic/yielding at concrete crushing; n: value not given in test report; –: no prestressing.
[b] upper values: $\lambda = 1$ (maximum diagonal crack spacing); lower values: $\lambda = 0.5$ (minimum diagonal crack spacing).

Tab. 5.4 – Orthotropically reinforced panels: summary of results.

crete while the x-reinforcement remained elastic. Failure modes are not indicated in the test report for the panels of Series HB; the remaining panels, including the partially prestressed specimens PP2 and PP3, failed by yielding of all reinforcement and subsequent crushing of the concrete. From Tab. 5.4 it can be seen that failure modes are correctly predicted for all but two specimens. The stresses in the reinforcement obtained at the ultimate state from the cracked membrane model are about 2.5 % above the yield stress for the x-reinforcement in Specimen PV10 and about 1 % below the yield stress for z-reinforcement in Specimen PV22, while in the experiment, the z-reinforcement yielded and the x-reinforcement was elastic in both panels. The differences between the predicted stresses at the ultimate state and the yield stress are very small and within the typical range of the scatter of yield strengths [65]; also, the determination of failure conditions is questionable unless steel strains are measured locally on the reinforcing bars. Hence, the prediction of failure modes is good, particularly if compared with other approaches. For example, Vecchio and Collins' [156] modified compression field approach results in stresses in the non-prestressed x-reinforcement of less than 75 % of the yield stress at the ultimate state for the partially prestressed specimens PP2 and PP3 [67], while these specimens failed by yielding of all reinforcement and subsequent crushing of the concrete.

As for the isotropically reinforced specimens, the cracked elastic response according to Eq. (5.17) and the ultimate load obtained from Eq. (5.33) closely approximate the actual behaviour of all specimens.

5.4   Additional Considerations

5.4.1   General Remarks

The different methods presented in Chapters 5.2 and 5.3 should be applied judiciously, depending on the nature of the problem under consideration and the required level of sophistication. Frequently, the determination of failure conditions according to limit analysis is sufficient; as illustrated in Figs. 5.9 to 5.12, ultimate loads can be accurately predicted by Eq. (5.33). Usually, such considerations are accompanied by an uncracked response analysis that allows one to judge the probability and the extent of cracking under service conditions. The approximate analytical solution derived in Chapter 5.2.4, Eq. (5.17), may be used for a simplified analysis of the cracked elastic response, either including or excluding tension stiffening effects. Figs. 5.9 to 5.12 demonstrate that the response obtained from the general numerical method is approximated well by the combined application of Eqs. (5.17) and (5.33).

5.4.2   Prestressing and Axial Stresses

The tension chord model, Chapter 2.4.2, can be extended to account for various types of reinforcement with different bond properties in the same direction, including combined prestressed and non-prestressed reinforcement as a special case; Alvarez [6] gives a detailed examination of the related problems. As a result, maximum stress-average strain relationships similar to Eqs. (2.19) to (2.21) are obtained for the different types of reinforcement. Noting that the average strain of each reinforcing bar is equal to the sum of its initial prestrain and the average strain of the tension chord, these relationships can be combined to a maximum stress-average strain relationship of the tension chord. By determining the reinforcement stresses from the resulting relationships in either reinforcement direction rather than from Eqs. (2.19) to (2.21), application of the cracked membrane model to orthogonally reinforced panels with reinforcement having different bond properties in the same direction, including prestressed reinforcement, is straightforward.

According to limit analysis ultimate loads are not affected by a possible prestressing of the reinforcement, and axial compressive stresses $-\sigma_x$ or $-\sigma_z$ are equivalent to $\rho_x f_{syx}$ and $\rho_z f_{syz}$, respectively, as long as the conditions $-\sigma_x \le f_c + \rho_x f'_{syx} - (\rho_z f_{syz} - \sigma_z)$ and $-\sigma_z \le f_c + \rho_z f'_{syz} - (\rho_x f_{syx} - \sigma_x)$ apply, see Chapter 5.3.2. In reality, prestressing and axial compressive stresses result in higher cracking loads, and if their direction coincides with that of the higher effective reinforcement ratio, Chapter 5.3.2, they reduce the reorientation of the internal forces after cracking since the initial cracks are oriented closer to the principal compressive stress direction at the ultimate state. This results in a delayed degradation of the concrete compressive strength and higher ultimate loads as confirmed by the panels of Series PP [91], Chapter 5.3.3. The non-prestressed panel PP1 failed by crushing of the concrete, while in the partially prestressed specimens PP2 and PP3, the concrete strength was sufficient to achieve failure by yielding of all reinforcement.

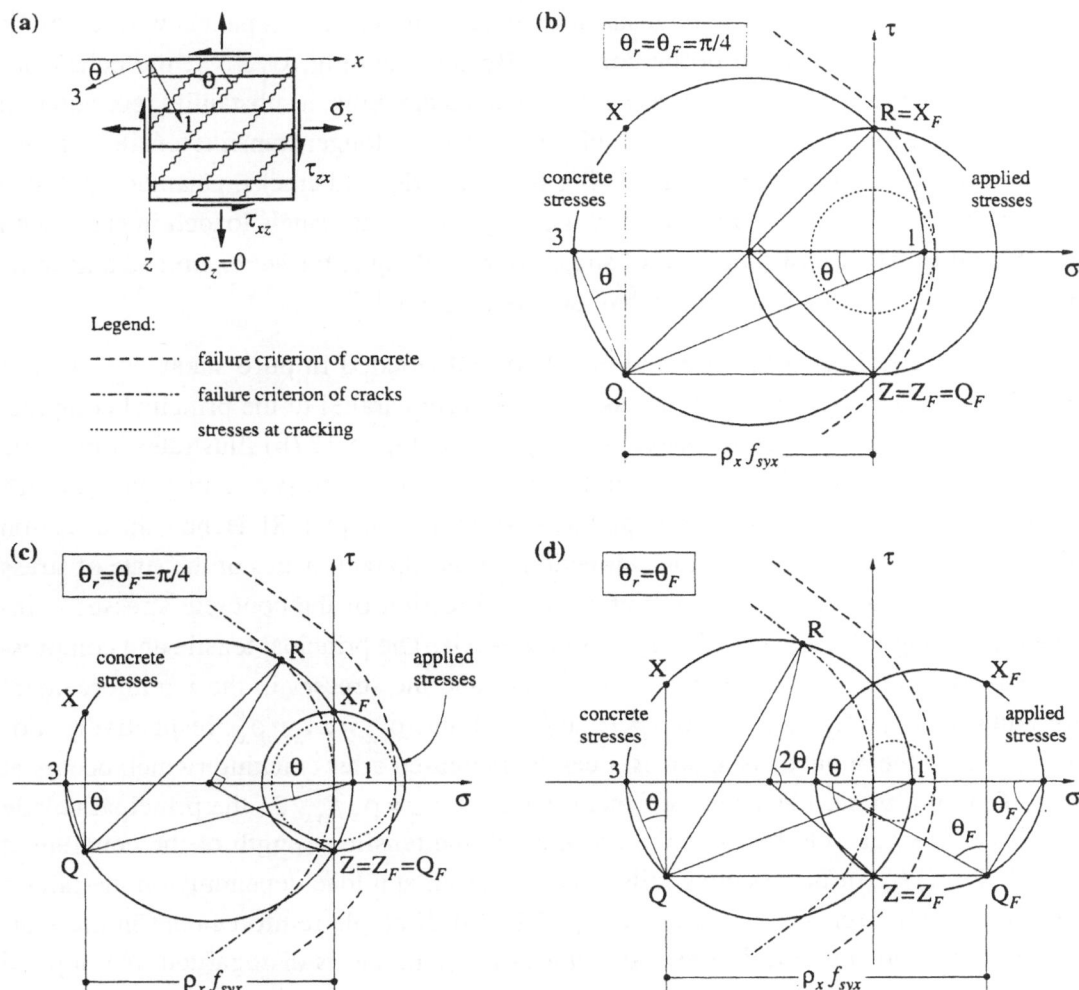

Fig. 5.13 – Uniaxially reinforced panels: (a) notation; (b) panel loaded in pure shear, cracks capable of transmitting shear stresses without accompanying normal stresses; (c) and (d) realistic failure criterion for the cracks, panels loaded in pure shear and in shear and axial tension, respectively.

5.4.3 Uniaxially Reinforced Elements

According to models considering rotating, stress-free cracks, panels reinforced only in the x-direction, Fig. 5.13 (a), are not capable of resisting shear stresses τ_{xz} unless compressive axial stresses $\sigma_z < 0$ are applied, see Eq. (5.2). Thus, for $\rho_z = 0$ and axial stresses $\sigma_z \geq 0$, zero shear resistance is predicted by the yield conditions for membrane elements according to limit analysis, Chapter 4.2.2, by the original compression field approaches, Chapter 4.3.2, and by the cracked membrane model, Chapter 5.2. Modified compression field approaches considering rotating cracks but expressing equilibrium in terms of average stresses between the cracks, Chapter 4.3.3, may predict shear resistances for uniaxially reinforced panels [32,14] even for $\sigma_z = 0$; this is due to the conceptual peculiarities of these approaches, see Chapters 4.3.3 and 5.3.1.

Shear stresses τ_{xz} can only be resisted by uniaxially reinforced panels with $\sigma_z \geq 0$ if tensile stresses are allowed in the concrete. Before cracking, $\sigma_1 \leq f_{ct}$, the behaviour closely matches that of plain concrete specimens. At cracking, a substantial reorientation of the internal forces takes place since the cracks can no longer transmit tensile stresses. Compression field approaches considering fixed, interlocked cracks, Chapter 4.3.3, can predict the post-cracking behaviour of uniaxially reinforced panels loaded in shear with $\rho_z = 0$ and $\sigma_z \geq 0$. Below, the state of stress after cracking is further examined and some basic aspects of the load-carrying behaviour are discussed.

Consider a uniaxially reinforced panel, $\rho_z = 0$, loaded in pure shear, $\sigma_x = 0$ and $\sigma_z = 0$, Fig. 5.13 (a). The cracks are assumed to form parallel to the principal compressive direction of the applied stresses, $\theta_r = \theta_F = \pi/4$. Fig. 5.13 (b) illustrates a possible state of stress after cracking, assuming that arbitrary shear stresses acting on the crack faces can be resisted without accompanying normal stresses [5,118]. Hence, the direction perpendicular to the cracks and the z-direction are the characteristics of the state of stress in the concrete, and the principal compressive direction of the concrete stresses is inclined at an angle $\theta = \theta_r/2$ with respect to the x-axis. The principal tensile and compressive stresses in the concrete between the cracks and the stresses in the x-reinforcement are equal to $\sigma_{c1} = \tau_{xz}(\sqrt{2}-1)$, $\sigma_{c3} = -\tau_{xz}(\sqrt{2}+1)$ and $\sigma_{sx} = 2\tau_{xz}/\rho_x$, respectively. For $\rho_x f_{syx} > 2 f_{ct}$ the applied shear stresses can be increased after cracking (which occurs at $\tau_{xz} = f_{ct}$) until either the x-reinforcement yields at $\tau_{xz} = \rho_x f_{syx}/2$, the principal tensile stresses in the concrete between the crack reach the tensile strength of the concrete at $\tau_{xz} = (\sqrt{2}+1)f_{ct}$, or the concrete fails in compression at a load depending on the failure criterion adopted for the concrete, see Fig. 5.13 (b). If ample reinforcement in the x-direction is provided and failure by crushing of the concrete is disregarded, the applied shear stresses can thus be increased by a factor of $(\sqrt{2}+1)$ after cracking.

In reality, only very small shear stresses can be transferred across cracks without accompanying normal stresses. If a realistic failure criterion for the cracks is adopted, Fig. 5.13 (c), much higher stresses are required in the x-reinforcement in order to transmit the stresses across the cracks, and the applied shear stresses can be only slightly increased after cracking. Hence, neglecting the normal stresses acting on the crack faces [5,118] is not only unrealistic, but also unconservative.

Cases with axial tensile stresses $\sigma_x > 0$ can be similarly treated, Fig. 5.13 (d). Assuming $\theta_r = \theta_F$, steeper cracks are obtained, $\theta_r > \pi/4$, and if the same crack failure criterion is used as for pure shear, the additional x-reinforcement (in addition to that resisting the axial tensile stresses, $\rho_x f_{syx} = \sigma_x$) required to transmit the applied shear stresses is even somewhat smaller than that for pure shear and $\theta_r = \pi/4$.

In summary, uniaxially reinforced panels with $\sigma_z \geq 0$ will not necessarily fail at initial cracking if ample x-reinforcement is provided. However, it should be emphasised that the load-carrying behaviour illustrated in Fig. 5.13 relies on the concrete tensile strength, that the shear resistance will break down if additional cracks form, and that existing, flat cracks may considerably reduce the shear strength of uniaxially reinforced panels.

6 Behaviour of Beams in Shear

6.1 General

The discontinuous stress fields outlined in Chapter 4.2 provide a powerful and efficient design tool for structural concrete girders. According to the lower-bound theorem of limit analysis, Chapter 3.2.2, these models indicate the necessary amount, the correct position and the required dimensioning and detailing of the longitudinal as well as the transverse reinforcement and result in safe designs since the flow of forces is followed consistently throughout the structure. In order to ensure attainment of the theoretically determined ultimate loads a certain redistribution of the internal forces is required. However, the theory of plasticity does not answer the related questions of the demand for and the supply of deformation capacity, particularly regarding the degradation of the concrete compressive strength in the web due to transverse tensile strains. More detailed investigations are required to settle these questions.

In situations where all static and geometric quantities vary only gradually along the girder axis, estimates of the overall load-deformation response of the web of structural concrete girders can be obtained from a single cracked membrane element, using the average values of the stresses and strains at mid-depth of the web at the section under consideration. A similar approach, neglecting tension stiffening effects, was applied in a recent re-evaluation of shear tests on structural concrete girders [77]. Since the stresses and strains in the web are non-homogeneous (even if all static and geometric quantities vary only gradually along the girder axis), the web has to be subdivided into several elements in order to obtain more realistic response predictions. Following similar lines as in the implementation of the modified compression field approach presented by Vecchio and Collins [156] into finite element programs [70,139,157], plane stress elements according to the cracked membrane model, Chapter 5, could be established and the load-deformation behaviour of the webs of structural concrete girders could be investigated by finite element analyses. Basically, plane stress elements which allow for a linearly varying state of strain would be more suitable than the current formulation of the cracked membrane model which assumes a homogeneous state of strain.

Rather than implementing the cracked membrane model into a finite element program, simplified models for girders with flanged cross-sections will be derived in the following. Chapter 6.2 examines situations where all static and geometric quantities vary only gradually along the girder axis, and in Chapter 6.3 discontinuity regions are discussed. Throughout these chapters, rotating, stress-free rather than fixed, interlocked cracks are considered, and the web thickness b_w is assumed to be constant.

6.2 Continuity Regions

6.2.1 Basic Considerations

In continuity regions all static and geometric quantities vary only gradually along the girder axis. Fig. 6.1 shows the principal compressive stress trajectories in the web of a girder in a continuity region, assuming a uniaxial compressive stress field in the concrete. If rotating, stress-free cracks with a vanishing spacing are considered, the principal compressive stress trajectories coincide with the cracks. In contrast to the sectional design approach presented in Chapter 4.2, the trajectories are not assumed to be straight and hence, the horizontal resultant of the diagonal compressive stresses in the concrete is not acting at mid-depth of the web but at a distance e_v above it, Fig. 6.1 (b). Rather than by Eq. (4.3)$_1$, the chord forces are thus determined by

$$F_{sup,inf} = \mp \frac{M}{d_v} + \frac{|V|}{2}\cot\theta\left(1 \pm \frac{2e_v}{d_v}\right) + \frac{N}{2} \tag{6.1}$$

provided that, as for Eq. (4.3), the sectional forces M and N are reduced to mid-depth of the web rather than the girder axis. Typically, the principal compressive stress direction is flatter at the compression chord, and the shear stresses are approximately constant over the depth of the cross-section, see Chapter 6.2.2. Hence, the tension chord force obtained from Eq. (6.1) is somewhat smaller than according to the sectional design approach, Eq. (4.3)$_1$, and the chord forces can be safely determined from Eq. (4.3)$_1$ in design practice. However, higher stirrup forces and higher principal concrete compressive stresses $-\sigma_{c3}$ are obtained near the tension and compression chords, respectively, if curved rather than straight trajectories are considered, see Fig. 4.2 (d) and Eq. (4.2). The principal concrete compressive stress trajectories are always curved if bonded reinforcement is considered, see Chapter 6.2.2, and in order to arrive at the uniform uniaxial compressive stress field with straight trajectories typically assumed in design practice, some redistribution of the internal forces is required.

Assuming that the state of stress in the web is constant in the horizontal direction, compression field approaches for continuity regions can be developed as outlined in Chapter 4.3.1. The axial strains are determined by the chord deformations if a certain strain variation over the depth of the cross-section is assumed, and the equilibrium condition in the axial direction is not involved in determining the states of stress and strain in the web. Equilibrium in the axial direction is satisfied integrally over the entire cross-section by determining the chord forces from Eq. (6.1); generally, an iterative procedure is required since e_v depends on the state of stress in the web obtained from the assumed variation of the axial strains. The concrete stresses in the web are equilibrated by stresses in the adjoining sections since the state of stress in the web is assumed to be constant along the girder axis. If no horizontal reinforcement is provided in the web and uniformly distributed shear stresses act on the web at the girder end, the end-sections of the girder have to resist the corresponding axial stresses.

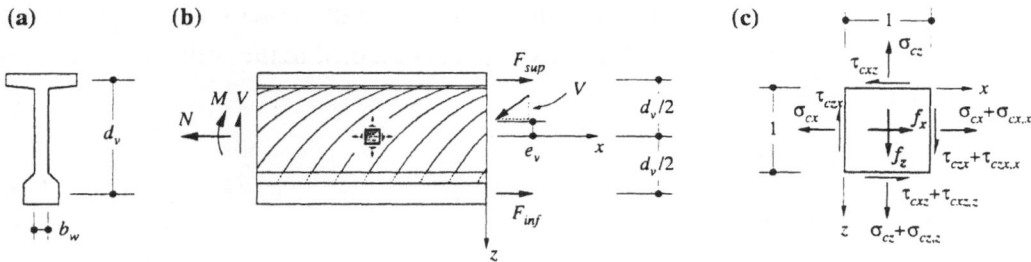

Fig. 6.1 – Continuity region of a girder: (a) cross-section; (b) elevation; (b) unit concrete element.

In principle, sectional models like the compression field approach outlined above are inappropriate if shear forces are present; the load-deformation behaviour always depends on the conditions in the neighbouring regions [45,54,110]. The assumption of constant stresses in the web along the girder axis is only approximately satisfied; due to the variation of the bending moment caused by the applied shear force the state of strain varies in the horizontal direction and hence, the stresses in the web cannot be constant. However, only moderate variations of the state of strain occur in continuity regions since by definition, they are situated away from the locations where concentrated loads causing peak moments and high chord strains are applied (peak moments in the span are irrelevant for the present thesis since $V = 0$ at such locations).

6.2.2 Stresses and Strains in the Web

Equilibrium of a unit element of the concrete in the web subjected to distributed forces (per unit area) f_x and f_z, Fig. 6.1 (c), requires that

$$\sigma_{cx,x} + \tau_{cxz,z} + f_x = 0$$

$$\tau_{cxz,x} + \sigma_{cz,z} + f_z = 0 \qquad (6.2)$$

where the comma designates partial derivatives with respect to the variable following the comma. The forces f_x and f_z comprise the forces transferred between the concrete and the reinforcement by bond shear stresses; if no horizontal web reinforcement is provided and no horizontal forces are applied in the web, one obtains $f_x = 0$. Assuming that the state of stress in the web is constant along the girder axis, $\sigma_{cx,x} = 0$ and $\tau_{cxz,x} = 0$, Eq. (6.2) yields $\sigma_{cz,z} + f_z = 0$ and $\tau_{cxz} =$ constant. Hence, noting that the integral of the shear stresses τ_{cxz} over the depth of the cross-section must be equal to the applied shear force V one obtains

$$\tau_{cxz} = \tau_{xz} = \frac{V}{b_w d_v} \qquad (6.3)$$

where τ_{xz} denotes the nominal applied shear stress.

Considering rotating, stress-free and orthogonally opening cracks with a vanishing spacing a uniaxial concrete compressive stress field is obtained in the web, and equilibrium requires that

$$\sigma_{sz} = (\sigma_z + \tau_{xz}\tan\theta)/\rho_z$$
$$\sigma_{c3} = -\tau_{xz}(\cot\theta + \tan\theta)$$

(6.4)

where θ denotes the inclination of the principal compressive stress (and strain) direction with respect to the x-axis, see Eqs. (4.9) and (5.2). If no vertical forces are applied in the web, one obtains $\sigma_z = $ constant, and noting that $-\sigma_{cz} = -\sigma_{c3}\sin^2\theta = \tau_{xz}\tan\theta$, differentiation of Eq. (6.4) with respect to z yields the expressions

$$-\sigma_{cz,z} = \tau_{xz}(\tan\theta)_{,z} \qquad\qquad \sigma_{sz,z} = \frac{\tau_{xz}(\tan\theta)_{,z}}{\rho_z}$$

(6.5)

for the vertical stresses in the concrete and in the stirrups. The forces $f_z = \tau_{xz}(\tan\theta)_{,z}$ are transferred between the concrete and the reinforcement by bond shear stresses, see Eq. (6.2)$_2$. From Eq. (6.5) it can be seen that if the inclination of the principal compressive stress direction with respect to the x-axis is constant over the depth of the cross-section, as assumed by sectional design approaches and parallel stress bands, Chapter 4.2, no stresses are transferred between the concrete and the reinforcement in the web.

This situation is illustrated in Fig. 6.2 (a), assuming a linear variation of the axial strains ε_x over the depth of the cross-section. The stresses and strains in the vertical reinforcement are constant, and, except for $z = -d_v/2$ and $z = d_v/2$, bond slip δ occurs, $\delta = u_s - u_c$, where u_s and u_c are the vertical displacements of the concrete and the stirrups, respectively, Chapter 2.4.1. The bond slip must vanish at the tension and compression chords and hence, the overall compatibility condition

$$\int_{z=-d_v/2}^{z=d_v/2} \varepsilon_{sz}dz = \int_{z=-d_v/2}^{z=d_v/2} \varepsilon_z dz$$

(6.6)

must be observed. The assumed linear variation of the axial strains ε_x results in linearly varying vertical strains $\varepsilon_z = \varepsilon_3 + (\varepsilon_x - \varepsilon_3)\cot^2\theta$ and shear strains $\gamma_{xz}/2 = (\varepsilon_x - \varepsilon_3)\cot\theta$ over the depth of the cross-section since θ and ε_3 are constant (if compression softening of the concrete due to $\varepsilon_1 = \varepsilon_x + (\varepsilon_x - \varepsilon_3)\cot^2\theta$ is accounted for, Chapter 2.4.3, ε_3 is no longer constant and the variations of ε_z, γ_{xz} become slightly non-linear). Thus, although the axial strains ε_x are linear in z, the cross-section will warp because of the linear variation of the shear strains over the depth of the cross-section [109,83]. A linear variation of ε_x must not be confused with the assumption made in beam theory that plane sections perpendicular to the girder axis remain plane. Generally, plane sections perpendicular to the beam axis remain plane if the deformations u in the direction of the girder axis are linearly distributed over the depth of the cross-section. If the axial strains $\varepsilon_x = u_{,x}$ are linear in z the deformations u may still depend on an arbitrary function of z alone. Thus, a linear (or constant) variation of ε_x over the depth of the cross-section is necessary, but not sufficient for plane sections perpendicular to the beam axis to remain plane.

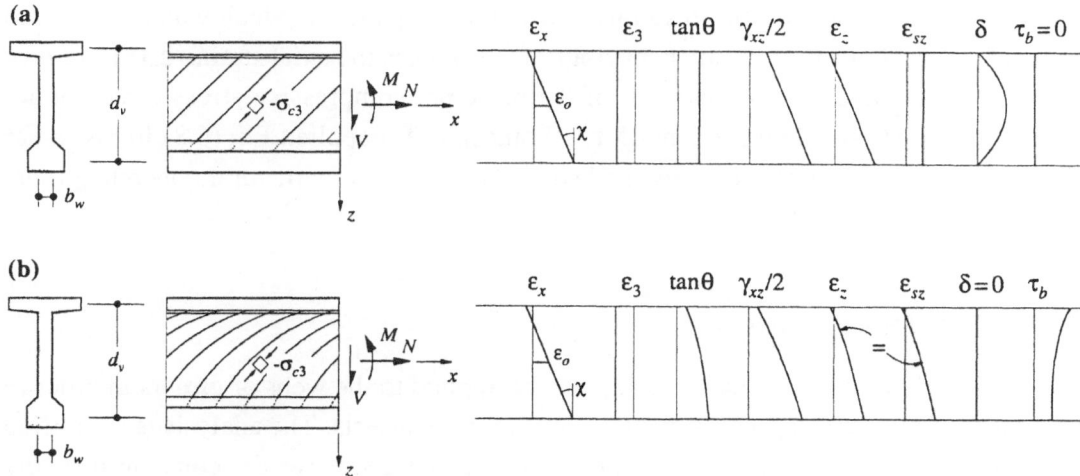

Fig. 6.2 – Continuity regions of girders, typical distributions of strains, bond shear stresses and bond slip: (a) unbonded reinforcement; (b) perfectly bonded reinforcement, cracks with a vanishing spacing.

If stresses are transferred between the concrete and the reinforcement in the web, the inclination of the principal compressive stress direction depends on z, Eq. (6.5), and the principal compressive stress trajectories are curved. Fig. 6.2 (b) illustrates a typical distribution of strains and bond shear stresses, assuming perfectly bonded reinforcement and orthogonally opening, rotating and stress-free cracks with a vanishing spacing. The principal compressive stress trajectories are affine since the state of stress in the web is assumed to be constant along the girder axis. The state of strain at any point of the web is completely determined by three non-collinear strains, Chapter 4.3.1. Thus, noting that $\sigma_z = 0$ and assuming a certain variation of the axial strains ε_x over the depth of the cross-section, the states of stress and strain in the web can be determined from Eq. (6.4) for given constitutive relationships of the concrete and the reinforcement. The stress-strain relationship of the naked steel can be applied for the stirrups since for perfectly bonded reinforcement the bond slip $\delta = u_s - u_c$ vanishes over the entire depth of the cross-section and hence, $\varepsilon_{sz} = \varepsilon_z$ is obtained. Knowing the stirrup stresses σ_{sz}, the corresponding bond shear stresses follow from Eqs. (6.5) and (2.12) as

$$\tau_b = \frac{\varnothing}{4s_{rm}}\sigma_{sz,z} = \frac{\tau_{cxz}\varnothing}{4s_{rm}\rho_z}(\tan\theta)_{,z} \qquad (6.7)$$

For linear elastic behaviour, Eq. (6.4) yields the expressions $\varepsilon_z = \tau_{xz}\tan\theta/(E_s\rho_z)$ and $\varepsilon_3 = -\tau_{cxz}(\cot\theta_r + \tan\theta)/E_c$, and, noting that $\cot^2\theta = (\varepsilon_z - \varepsilon_3)/(\varepsilon_x - \varepsilon_3)$ one obtains

$$\tan^2\theta(1 + n\rho_z) = \cot^2\theta n\rho_z + \cot\theta\frac{\rho_z E_s}{\tau_{xz}}\varepsilon_x(z) \qquad (6.8)$$

for the inclination θ of the principal compressive stress direction with respect to the x-axis, where $n = E_s/E_c$ = modular ratio. Substituting $\varepsilon_x = (\varepsilon_{x\,sup} + \varepsilon_{x\,inf})/2$ in Eq. (6.8), where $\varepsilon_{x sup}$, $\varepsilon_{x inf}$ are the chord strains, one obtains Eq. (4.12), which was first derived by Kupfer [72]. For linear elastic behaviour of the flanges the axial strains $\varepsilon_x(z)$ are proportional to τ_{xz} and the inclination θ of the principal compressive stress direction according to Eq. (6.8) is independent of τ_{xz}. Potucek [121] applied Eq. (6.8) in his linear elastic analysis of the states of stress and strain in the webs of structural concrete girders subjected to shear and flexure.

6.2.3 Finite Crack Spacings

The tension chord model, Chapter 2.4.2, can be applied to the webs of girders in order to account for finite crack spacings and tension stiffening effects. The analysis is simplified by the observation that according to Eq. (6.3), the shear stresses are constant over the depth of the cross-section if the state of stress in the web is assumed to be constant along the girder axis. However, the states of stress and strain in the web are non-homogeneous, and the sections at which the bond slip vanishes between consecutive cracks are not known beforehand.

If finite crack spacings are considered the assumption of rotating, stress-free and orthogonally opening cracks is – strictly speaking – incompatible with the curved cracks obtained from the variation of the state of strain over the depth of the cross-section, Chapter 6.2.2. Curved cracks open orthogonally over a finite distance only if the axial strains ε_t of the crack faces differ by $\Delta\varepsilon_t = \delta_n/r$, where δ_n = crack opening and r = radius of curvature of the crack. Apart from straight cracks, this condition will hardly ever be satisfied along an entire crack. Nevertheless, rotating, stress-free and orthogonally opening cracks are assumed here since the consideration of fixed, interlocked cracks would substantially complicate the investigations.

Consider an element of the web bounded by two consecutive cracks with inclinations θ_{rk} and $\theta_{r(k+1)}$, respectively, Fig. 6.3 (a). Knowing the stresses σ_{szrk} and $\sigma_{szr(k+1)}$ in the reinforcement at the cracks, the distributions of the bond shear and steel stresses, Fig. 6.3 (b), can be determined from Eq. (2.12), assuming a stepped rigid-perfectly plastic bond shear stress-slip relationship as in the tension chord model. At the location where the bond slip vanishes reinforcement stresses are minimal, and the concrete stresses $\Delta\sigma_{cz}$ reach their maximum value, Fig. 6.3 (b). The concrete stresses $\Delta\sigma_{cz}$, which have to be superimposed to the uniaxial diagonal compressive stress field in the concrete, are caused by the bond shear stresses τ_b and by the variation $\theta_{,z}$ of the principal compressive stress direction in the web, $\Delta\sigma_{cz} = \Delta\sigma_{cz}(\tau_b) + \Delta\sigma_{cz}(\theta_{,z})$. If no additional vertical forces are applied in the web, one obtains

$$-\Delta\sigma_{cz,z} = \frac{4\tau_b}{\varnothing}\frac{\rho_z}{(1-\rho_z)} - \tau_{xz}(\tan\theta)_{,z} \qquad\qquad \sigma_{sz,z} = \frac{4\tau_b}{\varnothing} \qquad (6.9)$$

The distributed forces $f_z(\theta_{,z}) = -\tau_{xz}(\tan\theta)_{,z}$ causing $\Delta\sigma_{cz}(\theta_{,z})$, see Fig. 6.3 (b), are the reactions to the change of the vertical component of the diagonal compressive stresses in the web, Eq. (6.5). They act on the concrete between the cracks and are transferred to the

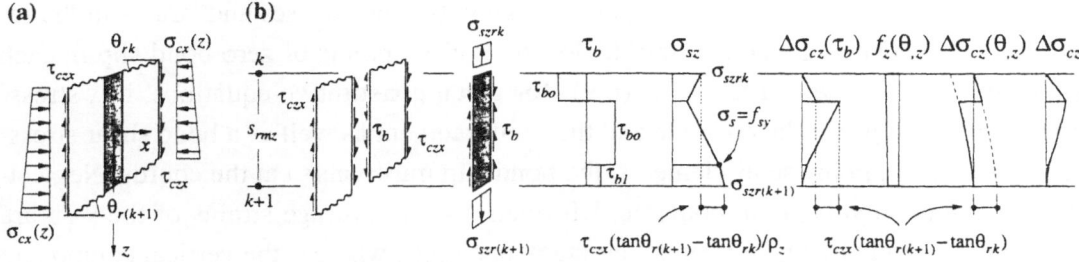

Fig. 6.3 – Bond behaviour of web element: (a) notation and applied forces; (b) vertical forces acting on concrete and reinforcement, respectively, and distribution of bond shear, steel and concrete stresses.

reinforcement via bond shear stresses. The integral of $f_z(\theta_{,z})$ over the element is equal to the (negative) difference of the vertical components σ_{czr} of the diagonal concrete compressive stresses at the cracks and it is also equal to the change of the stirrup stresses $\rho_z\sigma_{szr}$ caused by bond shear stresses, see Fig. 6.3 (b). Formulating equilibrium one obtains

$$\sigma_{szr(k+1)} - \sigma_{szrk} = \frac{\tau_{xz}(\tan\theta_{r(k+1)} - \tan\theta_{rk})}{\rho_z} \tag{6.10}$$

The condition $\sigma_z = 0$ is satisfied throughout the element since it is fulfilled at the cracks and $\sigma_{z,z} = (1-\rho_z)(\sigma_{cz,z} + \Delta\sigma_{cz,z}) + \rho_z\sigma_{sz,z} = 0$, see Eqs. (6.9) and (6.5)₁.

From Fig. 6.3 (b) it can be seen that the minimum steel stresses between the cracks are somewhat higher than if the element was symmetrically loaded, Chapter 2.4.2, while the maximum concrete stresses $\Delta\sigma_{cz}$ between the cracks are somewhat lower; the exact distribution of the concrete stresses $\Delta\sigma_{cz}$ depends on the variation of $f_z(\theta_{,z})$ over the element. Thus, compared to symmetrically loaded elements, tension stiffening effects are reduced since the bond shear stresses are partly used to transfer the vertical forces $f_z(\theta_{,z}) = -\tau_{xz}(\tan\theta)_{,z}$ from the concrete to the reinforcement. A detailed examination of tension stiffening with variable tensile forces, including a discussion of the influence of general bond shear-stress-slip relationships as well as suitable solution algorithms is given in a recent report by Alvarez [6].

If n_k cracks exist over the depth of the cross-section, the web can be subdivided into $(n_k - 1)$ elements like the one illustrated in Fig. 6.3 and two semi-elements next to the chords. The steel stresses σ_{szrk} at the cracks can be considered as the n_k primary unknowns in determining the states of stress and strain in the web. Knowing σ_{szrk} and noting that $\sigma_z = 0$, the principal concrete compressive stresses σ_{c3rk} at the cracks and the crack inclinations θ_{rk} follow from Eq. (6.4), assuming rotating, stress-free cracks. For a given distribution of the axial strains ε_x over the depth of the cross-section the state of strain at each crack can be determined from the stress-strain relationship of the concrete; the principal compressive strain direction at the cracks coincides with the crack direction since the cracks are stress-free and open orthogonally. The stresses and strains in the reinforcement between the cracks and the location of the points of zero bond slip in each element can be calculated from Eq. $(6.9)_2$ for given constitutive equations, i.e., stress-strain relationships for the concrete and the reinforcement as well as a bond shear stress-slip relationship; in the semi-elements, the bond slip must vanish at the chords. Neglecting the comparatively small concrete deformations, the average strains of the web are caused exclusively by the perpendicular crack openings, whereas the vertical component of each crack opening corresponds to the elongation of the reinforcement between the adjacent points of zero bond slip. The state of strain at each crack determined in the way outlined above can thus be attributed to the interval between the adjacent points of zero bond slip. Hence, compatibility requires that the integrals of the vertical strains ε_z and the steel strains ε_{sz} must be equal between any two adjacent points of zero bond slip; the strains ε_z correspond to the state of strain at the cracks and are constant over each interval. Since compatibility must be satisfied over each of the k intervals between two adjacent points of zero bond slip, n_k compatibility conditions are obtained, and the n_k primary unknowns σ_{szrk} can be determined as long as a point of zero bond slip exists between any two consecutive cracks. Note that cracks in those regions where bond slip occurs over several elements remain closed; they have to be eliminated from the calculations.

The assumption that the state of stress in the web is constant along the girder axis implies that the horizontal crack spacing s_{rmx} is constant over the depth of the cross-section. Knowing s_{rmx}, the vertical crack spacings follow from $s_{rmz} = s_{rmx}\tan\theta_r$, see Fig. 5.1 (c). Hence, tension stiffening effects are most pronounced near the tension chord where the cracks are steepest and the crack spacings are largest. At such locations, the crack spacing could be determined from Eq. (5.8) or its approximation, Eq. (5.12), observing that no tensile stresses are transferred to the concrete in the horizontal direction, $\lambda_x = 0$, since no horizontal web reinforcement is provided. However, very large crack spacings result from the small stirrup reinforcement ratios, and s_{rmx} is typically governed by the crack spacing of the tension chord.

6.2.4 Approximate Solutions

Assuming suitable constitutive relationships for the concrete and the reinforcing steel as well as an appropriate bond shear stress-slip relationship, the general procedure outlined in Chapter 6.2.3 could be implemented in a computer program. However, the solution is numerically intricate, particularly if overall slip occurs over several cracks. Here, further simplifications are introduced, and an approximate solution for low gradients of the stirrup stresses is established, assuming that the points of zero slip are located at the centres between the cracks. Thus, the steel stresses at the cracks are related to the average strains by the same stress-strain relationships as in the cracked membrane model, Eqs. (2.19) to (2.21). Obviously, the assumption of points of zero slip being located at the centres between the cracks is appropriate if the changes of the stirrup stresses $(\sigma_{szr(k+1)} - \sigma_{szrk})$ between two consecutive cracks are small. For a linear variation of the axial strains ε_x

over the depth of the cross-section the change of ε_x between two consecutive cracks is equal to $(\varepsilon_{x(k+1)} - \varepsilon_{xk}) = (\varepsilon_{x\,inf} - \varepsilon_{x\,sup})s_{rm}/d_v$, and since s_{rm} and $(\varepsilon_{x\,inf} - \varepsilon_{x\,sup})$ are independent of d_v, low gradients of the stirrup stresses $(\sigma_{szr(k+1)} - \sigma_{szrk})$ and good approximations are obtained for comparatively high girders. For $d_v/s_{rm} \to \infty$, the solution according to the general procedure outlined in Chapter 6.2.3 is obtained. The ratio of the nominal bond shear stresses (used to transfer the change of the stirrup stresses between consecutive cracks over a distance s_{rmz})

$$\tau_{b\,nom} = \frac{(\sigma_{szr(k+1)} - \sigma_{szrk})\varnothing}{4s_{rmz}} = \frac{\tau_{xz}(\tan\theta_{r(k+1)} - \tan\theta_{rk})\varnothing}{4\rho_z s_{rmz}} \tag{6.11}$$

to the maximum bond shear stresses τ_{bo} (τ_{b1} after onset of yielding) can be used as a measure of the appropriateness of the approximation made.

Fig. 6.4 illustrates the results of an analysis according to the approximate solution for low gradients of the stirrup stresses, assuming a linear variation of the axial strains ε_x over the depth of the cross-section and values of $d_v = 1.00$ m, $s_{rmx} = 0.20$ m, $\rho_z = 0.3\%$, $f_{sy} = 500$ MPa, $f_{su} = 625$ MPa, $E_s = 200$ GPa, $\varepsilon_{su} = 0.05$, $\varnothing_z = 16$ mm, $f_c' = 30$ MPa, $\varepsilon_{co} = 0.002$ and $\tau_{b1} = \tau_{bo}/2 = f_{ct} = 0.3(f_c')^{2/3}$ in MPa. As in the general numerical method of the cracked membrane model, the concrete stresses have been calculated from Eqs. (5.3) and (5.4), and the steel stresses have been determined from Eqs. (2.19) to (2.21). The chord strains $\varepsilon_{x\,sup}$, $\varepsilon_{x\,inf}$ have been assumed to vary linearly with the applied shear stress τ_{xz}, $\varepsilon_{x\,inf} = 0.002(\tau_{xz}/\tau_{u\,nom})$ and $\varepsilon_{x\,sup} = 0$. Here $\tau_{u\,nom} = 2.0\rho_z f_{syz}$ is the nominal shear resistance, determined from the sectional design approach, see Chapter 4.2, assuming that the stirrups yield and that $\cot\theta = 2$ at the ultimate state. The theoretically correct stepped variations of the stresses and strains resulting from the finite crack spacings have been substituted by continuous curves in Fig. 6.4; the plotted values apply for a crack at the corresponding location. Accordingly, the nominal bond shear stresses $\tau_{b\,nom}$ have been calculated from Eq. (6.7) rather than Eq. (6.11).

From Fig. 6.4 it can be seen that failure occurs at $\tau_{cxz} = 1.18\tau_{u\,nom}$ by crushing of the concrete while the stirrups yield over the entire depth of the cross-section. Although the principal concrete compressive stresses are highest at the compression chord, the concrete crushes at the tension chord where – due to the high transverse strains – compression softening is most pronounced. Like in the numerical example for membrane elements, Fig. 5.4, the concrete strain at the peak concrete compressive stress is somewhat lower than ε_{co} since near ε_{co}, the increase in the parabolic stress-strain curve of the concrete, Eq. (5.3), is compensated by the progressive reduction of the concrete compressive strength due to increasing ε_1, Eq. (5.4). Small values of $\tau_{b\,nom}$ and hence, good approximations of the response according to the general numerical procedure are obtained after the onset of yielding, and the initially curved cracks become almost straight at the ultimate state. The calculations illustrate that a state of stress very close to a uniform uniaxial compressive stress field with straight trajectories – as typically assumed in design practice, Chapter 4.2 – is approached at the ultimate state, even for small stirrup reinforcement ratios causing substantial redistribution of the internal forces.

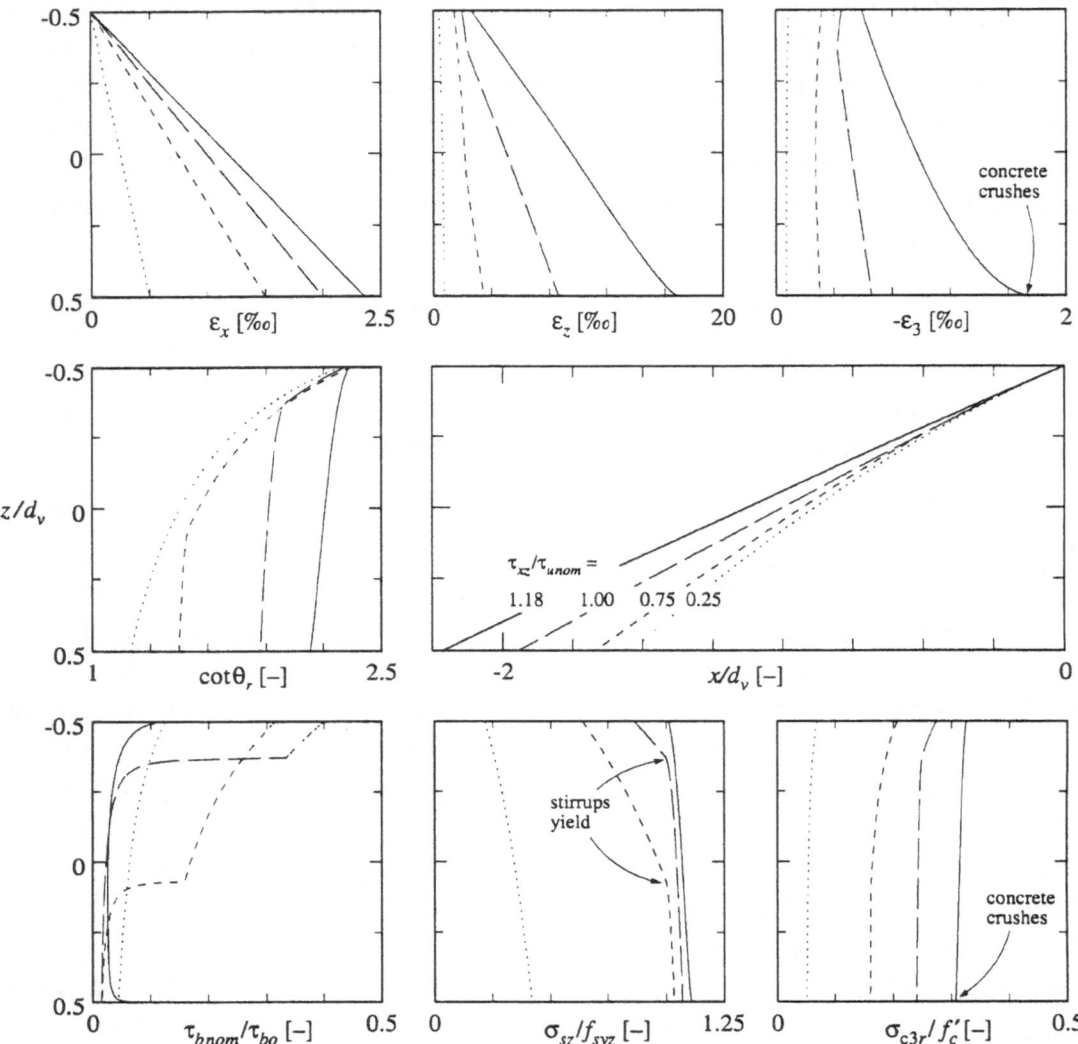

Fig. 6.4 – Continuity region: results of calculations according to the approximate solution for low gradients of the stirrup stresses.

The approximate solution for low gradients of the stirrup stresses is inadequate if bond slip occurs in the same direction over the entire depth of the cross section, similar to the situation illustrated in Fig. 6.2 (a). This is likely to occur if reinforcement with poor bond properties (plain bars) is used and in low girders with correspondingly high gradients of σ_z. Assuming a perfectly plastic bond shear stress-slip relationship, constant shear stresses τ_b act over the entire depth of the cross section and hence, one obtains

$$-\sigma_{cz,z} = \frac{4\tau_b}{\varnothing} \frac{\rho_z}{(1-\rho_z)} \tag{6.12}$$

Eq. (6.12) can be integrated analytically, and from $-\sigma_{cz} = -\sigma_{c3}\sin^2\theta = \tau_{xz}\tan\theta$, see Eq. (6.4)$_2$, one obtains $dz/dx = -\tan\theta = \sigma_{cz}/\tau_{cxz}$ and thus

$$\frac{dz}{dx} = -\frac{4}{\varnothing}\frac{\tau_b}{\tau_{cxz}}\frac{\rho_z}{(1-\rho_z)}z + \frac{\sigma_{cz}|_{z=0}}{\tau_{cxz}}$$ (6.13)

Integration yields the principal compressive stress trajectory passing through $x = z = 0$

$$z = e^{-\frac{4}{\varnothing}\frac{\tau_b}{\tau_{cxz}}\frac{\rho_z}{(1-\rho_z)}x + \ln\left(\tan\theta_o\frac{\varnothing\tau_{cxz}(1-\rho_z)}{4\ \tau_b}\frac{1}{\rho_z}\right)} - \tan\theta_o\frac{\varnothing\tau_{cxz}}{4}\frac{(1-\rho_z)}{\tau_b}\frac{1}{\rho_z}$$ (6.14)

where θ_o is the inclination of the trajectories at $z = 0$, to be determined such that overall compatibility is satisfied, Eq. (6.6).

Eq. (6.14) is of limited value; if reinforcement with poor bond properties is used, the trajectories according to Eq. (6.14) are almost straight, whereas for normal reinforcement, the assumption that bond slip occurs in the same direction over the entire depth is typically inappropriate after the onset of yielding, even for very low girders, since the gradients of σ_z become much lower once the stirrups yield, see Fig. 6.4.

6.2.5 Comparison with Experimental Evidence

Complete sets of strain readings in the web (three non-collinear strains at the same location) have rarely been recorded in tests on girders subjected to shear and flexure [77]. However, a comparison of the variations of stresses and strains obtained from the approximate solution for low gradients of the stirrup stresses with the experimental evidence requires two or more such sets of strain readings over the depth of the cross-section. According to the author's knowledge, the only experiments on profiled girders satisfying this condition are those by Kaufmann and Marti [65]; they involved two sets of strain readings in the web, using a 200 mm grid. Unfortunately, the crack spacing was of similar magnitude as the base length of the strain readings; the strains obtained from the 200 mm-grid are thus subject to considerable scatter and cannot be used for comparison purposes. Hence, the variations of stresses and strains over the depth of the cross-section cannot be verified since no pertinent experimental data is available. A comparison is only possible based on average strains measured over the entire depth of the web.

Fig. 6.6 compares the average strains measured in the web (using a 400 mm grid) in the experiments of Series VN by Kaufmann and Marti [65] with calculations according to the approximate solution for low gradients of the stirrup stresses. The experiments were conducted in the "Beam Element Tester" [65] of the Swiss Federal Institute of Technology, Zurich. This unique testing device, Fig. 6.5, is capable of subjecting elements of beams or columns with lengths of up to 6 m and cross-sectional dimensions of up to 0.80 m to arbitrary combined actions. Sectional forces corresponding to the ultimate resistance of heavily reinforced concrete girders (axial forces up to 5 MN and bending moments up to 3 MNm) can be applied at the element ends while maintaining a very close control of the deformations, allowing to perform precise large-scale experiments on beam elements rather than entire girders.

Basic experimental data and a summary of the conditions at the ultimate state are given in Tab. 6.1. Specimen geometry and reinforcement layout were the same for all tests (length 5.84 m, height 0.78 m, doubly symmetrical cross-section, web thickness $b_w = 0.15$ m, flange width 0.80 m, flange height 0.14 m, closed stirrups $\varnothing = 8$ mm spaced at 0.20 m). The only test parameter was the applied axial force N, see Tab. 6.1. In all experiments, axial forces (if any) were applied first; subsequently the shear force was gradually increased up to failure while the forces applied at the element ends were controlled such that axial forces remained constant throughout the test and bending moments vanished at midspan. The longitudinal reinforcement was chosen such that in all four tests, the chords remained elastic at the ultimate state.

The shear stresses indicated in Fig. 6.6 have been calculated from $\tau_{xz} = V/(b_w d_v)$, assuming that the effective shear depth is equal to the distance between the flange centres, $d_v = 0.64$ m. The experimental values of the strains shown in Fig. 6.6 are average values measured in the web at $x = \pm 1.0$ m from midspan, using a 400 mm grid. The calculations according to the approximate solution for low gradients of the stirrup stresses have been carried out at the same sections, assuming a linear variation of the axial strains ε_x over the depth of the cross-section. The values of the strains indicated in Fig. 6.6 are average values obtained in the range $z = -0.20 \ldots 0.20$ m, corresponding to the base length of the experimental strain readings. In all calculations, the chord strains $\varepsilon_{x\,sup}$, $\varepsilon_{x\,inf}$ have been assumed to vary linearly with the applied shear stresses τ_{xz} such that at the last load stage, the axial strains at $z = \pm 0.20$ m were equal to the experimentally observed values at the corresponding locations, Tab. 6.1. The stresses in the stirrups have been determined from Eqs. (2.19) to (2.21), assuming a horizontal crack spacing of $s_{rmx} = 150$ mm and $\tau_{b1} = \tau_{bo}/2 = f_{ct} = 0.3(f_c')^{2/3}$ in MPa, and the concrete stresses have been calculated from Eqs. (5.3) and (5.4).

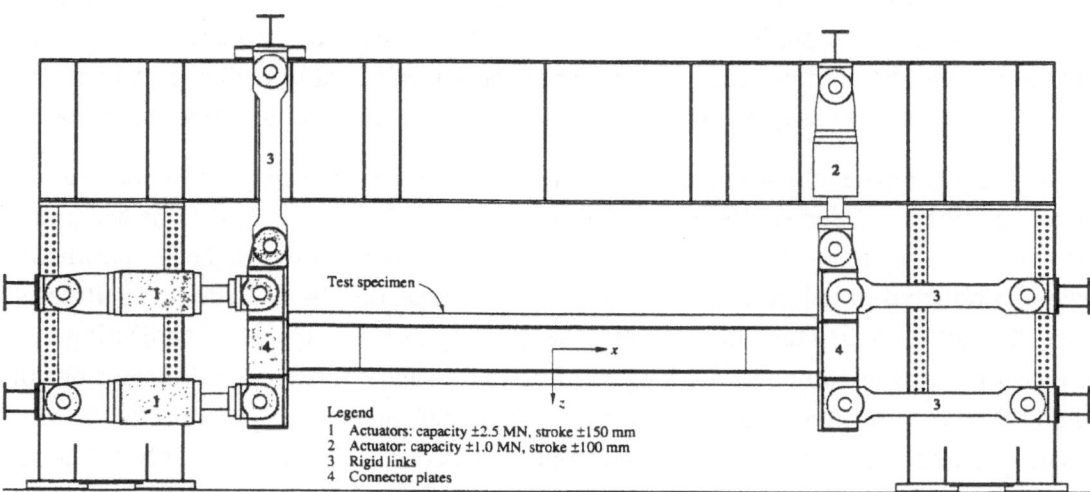

Fig. 6.5 – Experiments by Kaufmann and Marti [65]: Elevation of the Beam Element Tester, test setup for Series VN; scale 1:80.

Girder	VN1	VN2	VN3	VN4	Units
			Basic data		
N	0 [a]	0	1.00	-1.00	[MN]
ρ_z			0.34		[%]
$\varnothing_z$			8.0		[mm]
f_{syz}	546	511	511	511	[MPa]
f_{suz}	618	604	604	604	[MPa]
ε_{suz}	47.2	53.2	53.2	53.2	[‰]
E_s	203	210	210	210	[GPa]
f'_c	53.9	52.6	60.2	64.9	[MPa]
ε_{co}	2.56	2.36	2.27	2.40	[‰]
			Conditions at the ultimate state		
$\varepsilon_{xexp(z=-0.20\,m)}$	0.00	0.04	0.17	-0.03	[‰]
$\varepsilon_{xexp(z=+0.20\,m)}$	0.33	0.30	0.40	0.24	[‰]
τ_{uexp}	5.65	5.71	5.63	5.88	[‰]
τ_{uexp}/τ_{ucalc}	1.08	1.13	1.08	1.10	[–]
$\sigma_{sz(z=+0.20\,m)}$	601	565	570	568	[‰]

[a] specimen pre-cracked by axial tension and bending moments (constant over element) corresponding to maximum capacities of Beam Element Tester.

Tab. 6.1 – Basic data and summary of results of calculations illustrated in Fig. 6.6.

Generally, the agreement between the experimentally observed results and the calculations according to the approximate solution for low gradients of the stirrup stresses is very good. In all four experiments collapse was governed by failure of the web in the central portion of the specimens. For Specimens VN1, VN2 and VN3 collapse was triggered by stirrup rupture after some minor spalling of the web cover concrete, while Specimen VN4 exhibited a web crushing failure. According to the approximate solution for low gradients of the stirrup stresses, all four specimens failed by crushing of the concrete at $z = 0.20$ m after extensive yielding of the stirrups. If slightly higher values of the concrete compressive strength or the bond shear stresses had been adopted, stirrup ruptures would have been predicted. The calculated nominal shear stresses at the ultimate state are somewhat below the experimentally observed ultimate loads. Even lower ultimate loads would have been predicted if the calculations had been performed over the entire depth of the cross-section, accounting for the lower concrete compressive strengths at the tension chord according to Eq. (5.4). Thus, compression softening had a less pronounced effect on the concrete strength than predicted; this can be attributed to the fact that the webs of profiled girders are laterally restrained by the chords, resulting in an increase of the concrete compressive strength which counteracts the strength degradation by transverse tensile strains.

Ultimate loads could also be predicted as outlined in Chapter 5.3.2; assuming that $\varepsilon_{co} = 0.002$ and using the experimentally observed values of the axial strains ε_x at $z = 0.20$ m, Tab. 6.1, Eq. $(5.32)_2$ results in $\tau_{uexp}/\tau_{ucalc} = 1.14$, 1.20, 1.16 and 1.17 for Specimens VN1 to VN4, respectively. Slightly less conservative predictions are obtained if the axial strains ε_x at mid-depth of the web are used, i.e., $\tau_{uexp}/\tau_{ucalc} = 1.12$, 1.18, 1.14 and 1.15, respectively.

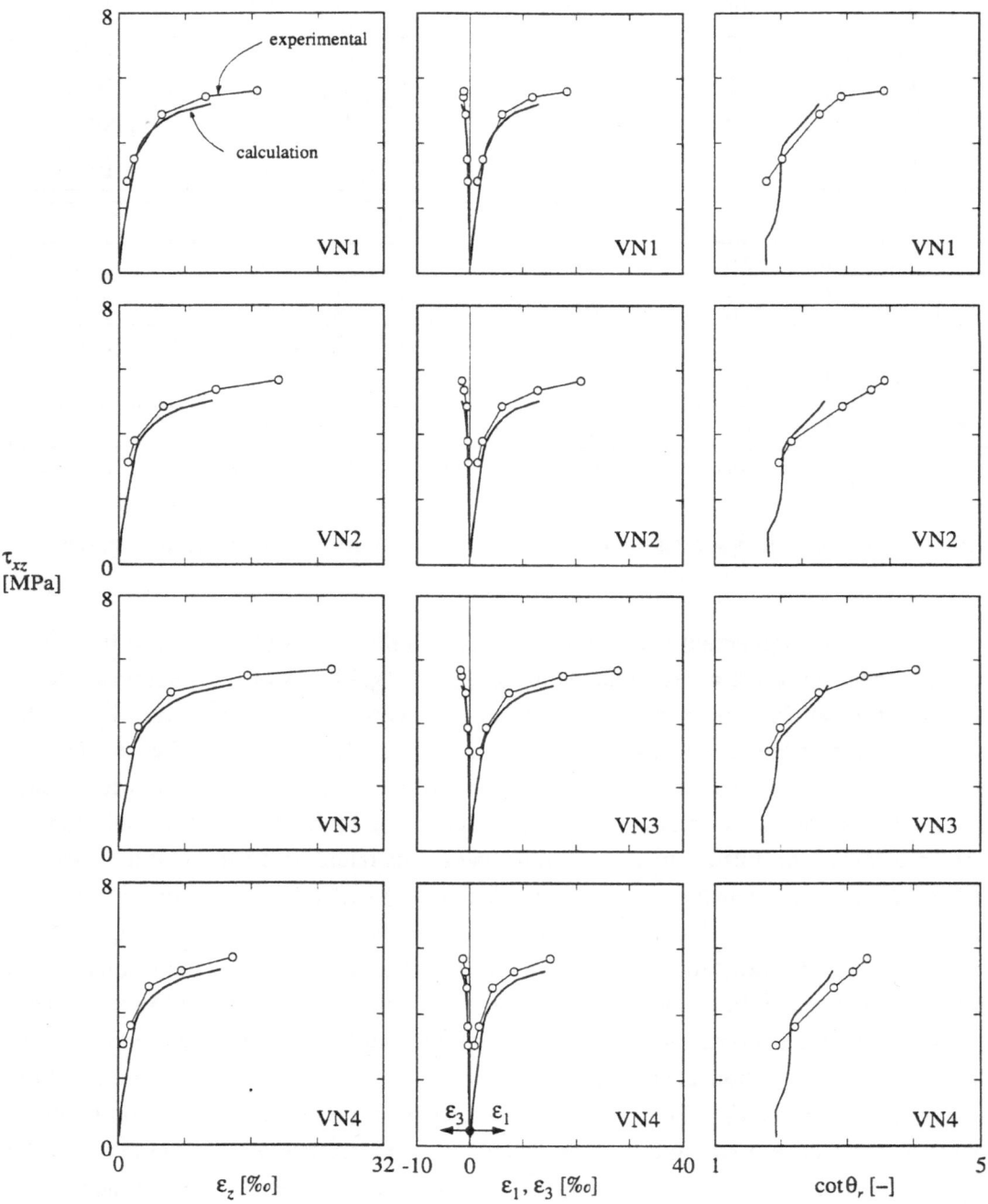

Fig. 6.6 – Approximate solution for low gradients of the stirrup stresses: Comparison with experiments of Series VN by Kaufmann and Marti [65].

6.3 Discontinuity Regions

6.3.1 Basic Considerations

Discontinuity regions are characterised by abrupt changes of static or geometric quantities. Generally speaking, the load-deformation behaviour of such regions is more complex than that of continuity regions. Here, only static discontinuities are further examined, using the practically relevant case of the support region of a constant-depth girder with flanged cross-section as an illustrative example. Fan-shaped discontinuous stress fields, Chapter 4.2.1, can be applied to visualise the force flow in such regions; a detailed examination of such stress fields is given in Chapter 6.3.2. Similar models can be developed for situations where the compression chord forces cannot be transferred to the outer regions of a flange (girders with rectangular cross-section, T-beams).

The effective concrete compressive strength f_c should be carefully selected in order to prevent brittle failure in support regions. For f_c = constant the concrete dimensions are governed by the nodal zone of the fan at the support where the concrete stresses are highest. If the influence of the state of strain on f_c is accounted for, see Eq. (5.4), the situation becomes more complex. Near the nodal zone where the concrete stresses are highest, f_c is enhanced by compressive axial strains, whereas axial tensile strains at the tension chord may markedly reduce f_c. In particular, if moment redistribution occurs at intermediate supports, plastic strains of the tension chord spread out to the regions where the flat trajectories of the fan, which undergo high compressive stresses, Eq. (4.2), join the tension chord, and crushing of the web near the tension chord may become critical. Currently, there are no methods available that allow checking whether the concrete stresses are below f_c throughout the fan region, apart from the unrealistic assumption that f_c = constant. In design practice, the web thickness b_w can be determined from the principal concrete compressive stresses in the adjoining parallel stress band [143,144], using a reasonably conservative value of f_c, such as Eqs. (2.25) or (2.26), in order to prevent potential web-crushing failures. The dimensions of the support plate are determined separately, assuming a much higher value of the concrete compressive strength. While this design procedure for b_w and the support dimensions can be justified by experimental evidence [21,142,143], it is not entirely satisfactory.

Chapter 6.3.3 presents a method that allows overcoming these difficulties. Starting from an estimate of the state of strain in the fan region, the effective concrete compressive strength predicted by Eq. (5.4) can be compared with the principal concrete compressive stresses in the fan region. Some numerical examples are given in Chapter 6.3.4, and Chapter 6.3.5 presents a brief comparison with the experimental evidence.

Basically, the load-deformation behaviour of support regions could be analysed using the model presented in Chapter 6.3.3. However, since the underlying estimate of the state of strain in the fan region is rather crude, the results of such load-deformation analyses would be inherently debatable and hence, no such analyses have been carried out.

6.3.2 Non-Centred Fans with Variable Concrete Strength

The force flow in support regions of girders with flanged cross-section can be modelled by fan-shaped stress fields, Fig. 4.2 (e) to (g). As outlined in Chapter 4.2.1, centred fans, Fig. 4.2 (e), are not suitable for this purpose owing to the strong concentration of concrete stresses in the flattest trajectory, whereas non-centred fans without a nodal zone, Fig. 4.2 (g), are of limited practical value since they require longer supports than fans with a nodal zone, Fig. 4.2 (f). Hence, only non-centred fans with a nodal zone are further examined below.

Consider the non-centred fan in the support region of the girder with flanged cross-section shown in Fig. 6.7. The stirrup forces f_w and the applied loads q (per unit length) as well as the tension and compression chord forces F_{supo} and F_{info} are assumed to be known. The flattest trajectory of the fan is inclined at an angle θ_o, corresponding to the inclination of the adjoining parallel stress band. The compression chord force F_{info} is

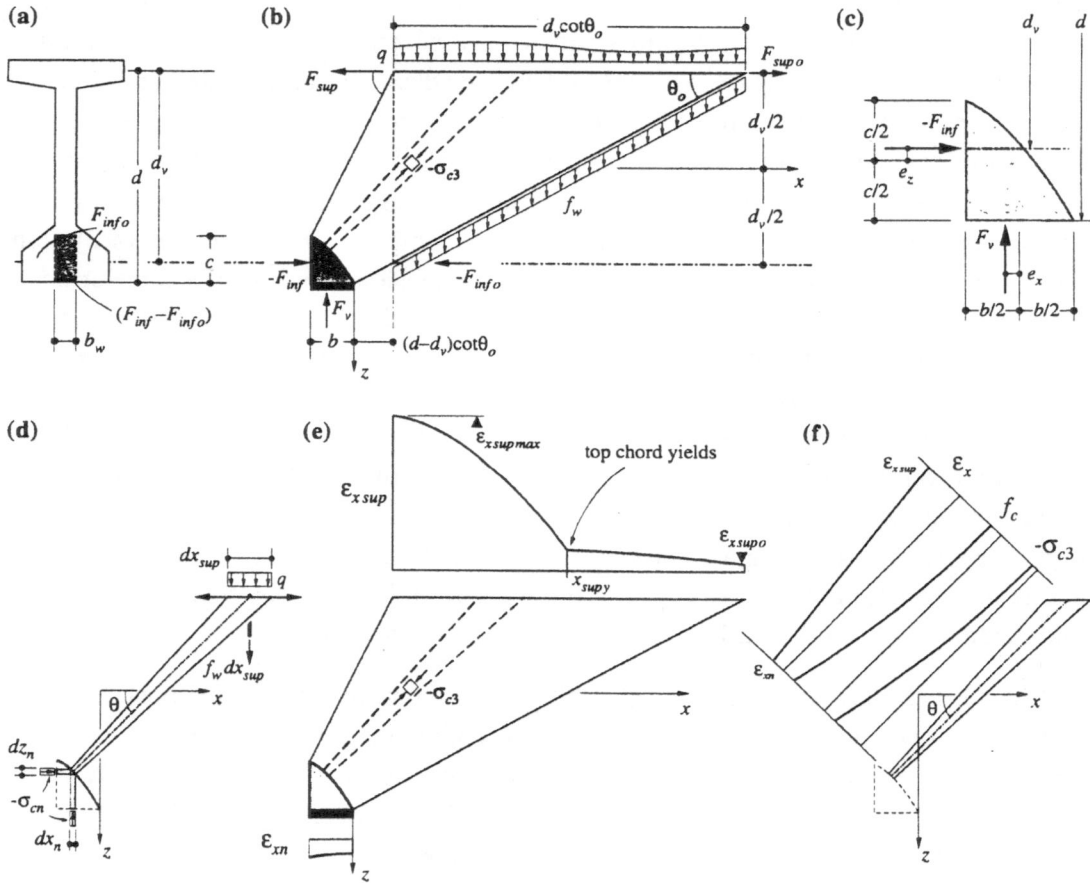

Fig. 6.7 – Discontinuity region, non-centred fan: (a) cross-section; (b) elevation; (c) detail of nodal zone; (d) infinitesimal strut; (e) variation of chord strains; (f) variation of axial strains, principal compressive stresses and effective concrete compressive strength along principal compressive stress trajectory.

assumed to be transferred to the outer regions of the flange and hence, only the horizontal resultant of the stresses carried by the fan, $(F_{inf}-F_{info})$, acts on the vertical boundary of the nodal zone of width b_w, Fig. 6.7 (a). The depth d and the effective shear depth d_v are assumed to be constant and have been chosen such that the horizontal force acting on the nodal zone $(F_{inf}-F_{info})$ acts at the same height $(z=d_v/2)$ as the compression chord force F_{info}, Fig. 6.7 (c). Generally, d or d_v must be adapted in an iterative manner in order to achieve this collinearity. The state of stress inside the nodal zone is not necessarily uniform, resulting in eccentricities e_x and e_z of the support reaction F_v and the horizontal force $(F_{inf}-F_{info})$ acting on the nodal zone, respectively, see Fig. 6.7 (c).

Formulating equilibrium of an infinitesimal strut of the fan, Fig. 6.7 (d), one obtains

$$\frac{dx_{sup}}{dx_n} = \frac{-b_w\sigma_{cn}(x_n)}{q(x_{sup})+f_w(x_{sup})} \tag{6.15}$$

where $\sigma_{cn}(x_n)$ are the principal concrete compressive stresses at the bottom end of the strut. The horizontal coordinates x_n, x_{sup} of the lower and upper strut end are related by

$$x_{sup} = x_n + (z_n+d_v/2)\cot\theta \tag{6.16}$$

where z_n = vertical coordinate of the boundary of the nodal zone, Fig. 6.7 (d). Differentiating Eq. (6.16) with respect to x_n one obtains

$$\frac{dx_{sup}}{dx_n} = 1 + \frac{dz_n}{dx_n}\cot\theta(x_n) + (z_n+d_v/2)\frac{d}{dx_n}[\cot\theta(x_n)] \tag{6.17}$$

The horizontal and vertical stresses inside the nodal zone are assumed to be equal at each point along the boundary to the fan, $\sigma_{cx}(x_n) = \sigma_{cz}(x_n) = \sigma_{cn}(x_n)$, Fig. 6.7 (d). Thus, the boundary of the nodal zone is perpendicular to the principal compressive stress direction in the fan, $\cot\theta = dz_n/dx_n$, and from Eqs. (6.15) and (6.17) one obtains the second-order differential equation for the boundary between the nodal zone and the fan

$$\frac{d^2z_n}{dx_n^2} = \frac{1}{(z_n+d_v/2)}\left[\frac{-b_w\sigma_{cn}(x_n)}{q(x_{sup})+f_w(x_{sup})} - 1 - \left(\frac{dz_n}{dx_n}\right)^2\right] \tag{6.18}$$

Using suitable numerical techniques, Eq. (6.18) can be integrated for arbitrary distributions of $\sigma_{cn}(x_n)$, $q(x_{sup})$ and $f_w(x_{sup})$, starting from $x_{no} = 0$ where the initial conditions $z_{no} = d-d_v/2$ and $(dx_n/dz_n)_o = \cot\theta_o$ have to be satisfied and noting that x_{sup} follows from Eq. (6.16) for known values (x_n, z_n). The end of the interval of integration, $x_{ne} = -b$, is not known beforehand but has to be determined iteratively from the condition that $x_{sup} = (d-d_v)\cot\theta_o$ must be obtained at $x_n = x_{ne}$.

Equilibrium of an infinitesimal strut of the fan, Fig. 6.7 (d), requires that the tension chord force varies in proportion with $b_w\sigma_{cn}z_n$; alternatively, the variation of the tension chord force can be determined from Eq. (4.1)$_1$, $dF_{sup}/dx_{sup} = -(q+f_w)\cot\theta$. According to Chapter 3.3.4 the principal compressive stress trajectories are the characteristics of the state of stress if the concrete does not crush. Thus, the principal concrete compressive stresses $\sigma_{c3}(z)$ vary hyperbolically along the trajectories of the fan

$$-b_w\sigma_{c3}(z) = \frac{(q+f_w)(1+\cot^2\theta)}{1-(z+d_v/2)\dfrac{d}{dx_{sup}}\cot\theta} = \frac{-b_w\sigma_{cn}}{1+(z_n-z)\dfrac{d^2z_n/dx_n^2}{1+(dz_n/dx_n)^2}} \tag{6.19}$$

see Eq. (3.17) and Fig. 6.7 (f). Eq. (6.19) allows determining the value of σ_{c3} throughout the fan region after the integration of Eq. (6.18). Obviously, the highest compressive stresses are obtained at the boundary to the nodal zone where $\sigma_{c3} = \sigma_{cn}$.

For $(q+f_w)$ = constant the support reaction F_v and the horizontal force $(F_{inf}-F_{info})$ acting on the nodal zone are equal to

$$F_v = (q+f_w)d_v\cot\theta_o$$
$$F_{inf}-F_{info} = -\frac{(q+f_w)d_v\cot\theta_o}{2}(b+2e_x+[2d-d_v]\cot\theta_o) \tag{6.20}$$

where e_x is the eccentricity of F_v with respect to the support, Fig. 6.7 (c).

For σ_{cn} = constant, the state of stress inside the nodal zone is uniform, and one may assume $\sigma_{cn} = -f_c$, corresponding to $\sigma_{c1} = \sigma_{c3} = -f_c$ inside the nodal zone. The forces F_v and $(F_{inf}-F_{info})$ act at the centres of the corresponding boundaries of the nodal zone, $e_x = e_z = 0$ and $c = 2(d-d_v)$, see Fig. 6.7 (c).

If both $\sigma_{cn} = -f_c$ = constant as well as $(q+f_w)$ = constant, the quantities b and c are equal to

$$b = \frac{F_v}{b_w f_c} = \frac{q+f_w}{b_w f_c}d_v\cot\theta_o$$
$$c = -\frac{F_{inf}-F_{info}}{b_w f_c} = \frac{b^2+bd_v\cot\theta_o}{2d_v-b\cot\theta_o} \tag{6.21}$$

For $\sigma_{cn}/(q+f_w)$ = constant, Eq. (6.18) can be solved analytically [82,143]. Noting that $z_{no} = d-d_v/2$ and $(dx_n/dz_n)_o = \cot\theta_o$ at $x_n = 0$, one obtains the expression

$$z_n = \sqrt{\left(\frac{b_w\sigma_{cn}}{q+f_w}-1\right)x_n^2+2d\cot\theta_o x_n+d^2} - \frac{d_v}{2} \tag{6.22}$$

for the boundary between the nodal zone and the fan.

In design practice it is generally not necessary to determine the variation of the chord forces or the geometry of the boundary of the nodal zone in detail. Typically, both $\sigma_{cn} = -f_c$ as well as $(q+f_w)$ are assumed to be constant, and the tension chord reinforcement as well as the transverse reinforcement can be determined from the simpler fan centred at $x = -b/2$, $z = d_v/2$, see Chapter 4.2.1. The determination of the web thickness b_w and the support dimensions b, c is further examined in the following chapters.

6.3.3 Stresses and Strains in the Fan Region

Knowing the effective concrete compressive strength $f_c(x_n)$ along the boundary of the nodal zone to the fan, Eq. (6.18) can be integrated numerically for any distribution of $q(x_{sup})$ and $f_w(x_{sup})$, assuming that the geometry of the fan is determined by the condition $\sigma_{cn}(x_n) = -f_c(x_n)$ at the boundary of the nodal zone. This assumption implies that the principal compressive strain ε_3 along the boundary of the nodal zone is approximately equal to $-\varepsilon_{co}$. Hence, if the axial strains ε_{xn} along the boundary are given, the value of $f_c(x_n)$ can be determined from Eq. (5.4), noting that $\varepsilon_1 = \varepsilon_x + (\varepsilon_x - \varepsilon_3)\cot^2\theta$ and that – for coinciding principal compressive stress and strain directions – the principal compressive strain direction is known at each step of integration, $\cot\theta = dz_n/dx_n$. The axial strains ε_{xn} along the boundary of the nodal zone cannot be determined from the discontinuous stress field, and strain-compatibility analyses for $x_n =$ constant are inappropriate in discontinuity regions since the experimentally observed variation of ε_{xn} over the depth of the cross-section is highly non-linear [142,65]. Hence, the axial strains along the boundary of the nodal zone, which are approximately equal to the axial strains of the compression chord, $\varepsilon_{xn} \approx \varepsilon_{x\,inf}$, will be used as a parameter in the calculations, assuming that $\varepsilon_{xn} =$ constant along the girder axis.

In order to check whether the concrete stresses are below the effective concrete compressive strength f_c according to Eq. (5.4) throughout the fan region, the state of strain in the fan must be known. Considering rotating, stress-free and orthogonally opening cracks with a vanishing spacing, the principal compressive direction of the strains coincides with the principal compressive stress trajectories of the fan which, according to Chapter 3.3.4, are the characteristics of the state of stress if the concrete does not crush, $\varphi = \pi/2$. They are straight since no distributed forces act on the concrete and, for deformations at the ultimate state, they remain straight and unstrained. Thus, the plastic strain increments vary linearly along the principal compressive stress trajectories. Assuming that the elastic portions of the strains are also linearly distributed along the principal compressive stress trajectories, the axial strains along the trajectories follow from the strains $\varepsilon_{x\,sup}$ of the tension chord and the axial strains ε_{xn} along the boundary of the nodal zone. Hence, assuming a certain value of the principal compressive strain ε_3, the state of strain and the effective concrete compressive strength f_c according to Eq. (5.4) in the fan region can be determined.

Assuming that ε_3 is constant in the fan region, $\varepsilon_3 \approx -\varepsilon_{co}$, one obtains a linear variation of the principal tensile strains $\varepsilon_1 = \varepsilon_x + (\varepsilon_x - \varepsilon_3)\cot^2\theta$ along the principal compressive stress trajectories, Fig. 6.7 (f), since the axial strains ε_x vary linearly and $\theta =$ constant. Hence, the effective concrete compressive strength according to Eq. (5.4) varies hyperbolically along the principal compressive stress trajectories unless the limitation $f_c \leq f_c'$ is governing. According to Eq. (6.19) the variation of the principal concrete compressive stresses $-\sigma_{c3}$ along the trajectories is also hyperbolic, Fig. 6.7 (f). Two hyperbolas $(c_1 + c_2 z)^{-1}$ intersect at no more than one point, and the geometry of the fan has been determined from the condition $-\sigma_{c3} = f_c$ at the lower end of all trajectories. Thus, the concrete stresses are below f_c throughout the fan region if they are below f_c along the tension chord. If the limitation $f_c \leq f'_c$ is governing at either end of a principal compressive stress trajectory, the actual value of f_c along the entire trajectory exceeds that obtained by assuming a hyperbolic variation $(c_1 + c_2 z)^{-1}$ between the values at the top and bottom ends. Hence, it can be concluded that if the concrete stresses are equal to f_c at the boundary of the nodal zone and equal to or below f_c along the entire tension chord, the concrete stresses are below f_c throughout the fan region.

The assumption $-\sigma_{c3} = f_c$ at the lower end of all trajectories results in the highest possible values of f_c along the tension chord. Concrete stresses below f_c at the nodal zone would result in lower gradients of $\cot\theta = dz_n / dx_n$, Eq. (6.18), flatter inclinations θ and higher values of $\varepsilon_1 = \varepsilon_x + (\varepsilon_x - \varepsilon_3)\cot^2\theta$ at the tension chord. The loads $q(x_{sup})$ and $f_w(x_{sup})$ are independent of the fan geometry if $\cot\theta_o$ is given and hence, if the concrete stresses exceed f_c somewhere along the tension chord, the loads $q(x_{sup})$ and $f_w(x_{sup})$ cannot be carried by a fan with the given web thickness b_w and inclination $\cot\theta_o$; the concrete will crush at the tension chord even if ample support width is provided. Thus, while the tension chord strains have no direct influence on the fan geometry determined from $-\sigma_{c3} = f_c(x_n)$ along the boundary of the nodal zone, high axial strains $\varepsilon_{x\,sup}$ impose limitations on the combinations of the parameters q, f_w, b_w, and $\cot\theta_o$ for which, according to the assumptions made, a solution is possible.

Often, the compression chord is very stiff and acts as an extension of the support plate, resulting in steeper trajectories, lower compressive stresses $-\sigma_{c3}$ and higher values of f_c along the tension chord. In such cases, the loads $q(x_{sup})$ and $f_w(x_{sup})$ can be increased until the compression chord is sheared off or the concrete crushes in the adjacent parallel stress band. Similar observations apply if the provided support width is smaller than required by a fan determined from the assumption $-\sigma_{c3} = f_c$ at the lower end of all trajectories. Eventually, failure will again occur through shearing off of the compression chord or through concrete crushing in the adjacent parallel stress band.

6.3.4 Numerical Examples

Knowing the chord forces according to the discontinuous stress field solution presented in Chapter 6.3.2 the axial strains ε_x along the tension and compression chords can be determined, using suitable constitutive relationships for the chords. Sigrist [144] applied this approach to problems of the deformation capacity of structural concrete, determining the chord forces from a fan with f_c = constant and using the tension chord model, Chapter 2.4.2. He obtained very good correlations with the experimentally observed chord strains and with the overall deformations [142].

Rather than determining the chord strains ε_x for a specific reinforcement layout, typical distributions of the tension chord strains $\varepsilon_{x\,sup}$ are assumed here for comparison purposes, whereas constant values of the axial strains $\varepsilon_{x\,n} \approx \varepsilon_{x\,inf}$ along the boundary of the nodal zone will be used as a parameter in the calculations. Hence, possible interactions between the fan geometry and the chord strains are neglected.

Typically, the tension chord strains $\varepsilon_{x\,sup}$ increase drastically in the regions where plastic strains occur, and distributions similar to that illustrated in Fig. 6.7 (e) are observed in tests [142,65]. The extent of the region characterised by plastic chord strains and the value of the maximum tension chord strain $\varepsilon_{x\,sup\,max}$ depend on the type and layout of the reinforcement as well as on the amount of moment redistribution at intermediate supports; values of $\varepsilon_{x\,sup\,max} = 10...40\%_o$ have been observed in tests [142,65]. A simple approximation of the actual distributions of the tension chord strains will be adopted for the calculations carried out in this chapter, assuming a constant value, $\varepsilon_{x\,sup} = \varepsilon_{x\,sup\,o}$, in the regions where no plastic chord strains occur, $x_{sup} > x_{sup\,y}$, and a parabolic increase from $\varepsilon_{x\,sup\,o}$ to $\varepsilon_{x\,sup\,max}$ for $x_{sup} < x_{sup\,y}$, i.e.

$$\varepsilon_{x\,sup} = \varepsilon_{x\,sup\,o} + (\varepsilon_{x\,sup\,max} - \varepsilon_{x\,sup\,o})[1 - (\xi/\xi_y)^2] \qquad \text{for } \xi < \xi_y \qquad (6.23)$$

where the relative coordinate $\xi = 0...1$ and the parameter ξ_y, measuring the extent of the region characterised by plastic chord strains, are determined by

$$\xi = \frac{x_{sup} - (d - d_v)\cot\theta_o}{d_v \cot\theta_o} \qquad\qquad \xi_y = \frac{x_{sup\,y} - (d - d_v)\cot\theta_o}{d_v \cot\theta_o} \qquad (6.24)$$

According to Eq. (4.2), the principal compressive stresses in the flat trajectories of the fan, joining the tension chord at comparatively high values of ξ, are much higher than the stresses in the steep trajectories. Hence, ξ_y has a more pronounced effect on the combinations of the parameters q, f_w, b_w, and $\cot\theta_o$ for which a solution is possible without crushing of the concrete at the tension chord than the value of $\varepsilon_{x\,sup\,max}$. Thus, a constant value of $\varepsilon_{x\,sup\,max} = 20\%_o$ is assumed in all numerical examples, while the influence of ξ_y is further examined.

Fig. 6.8 illustrates the results of an example calculation using a non-centred fan with f_c according to Eq. (5.4). The geometry of the fan has been calculated from integrating Eq. (6.18), using a fourth order Runge-Kutta solution algorithm and assuming that $-\sigma_{c3} = f_c(x_n)$ along the boundary of the nodal zone to the fan. The loads q and f_w have been assumed to be constant, $(q + f_w) = F_v/(d_v \cot\theta_o)$, and the support reaction F_v has been chosen such that for the given values of b_w and $\cot\theta_o$, the concrete in the adjoining parallel stress band is at the onset of crushing, $-\sigma_{c3} = f_c$ for $z = -d_v/2$ and $\xi = 1$. For comparison purposes the results of calculations using a non-centred fan with constant f_c, Eqs. (6.21) and (6.22), are also shown. The average value of f_c has been chosen such that the same support width b results from Eq. (6.21)$_1$ as for the solution with variable f_c. Basic values underlying the calculations are given in Fig. 6.8.

From Fig. 6.8 it can be seen that a support width of $b = 184$ mm is required and that the effective depth must not be less than $d = 1.134$ m in order to achieve that the horizontal force $(F_{inf} - F_{inf\,o})$ acts at the same height $(z = d_v/2)$ as the compression chord force $F_{inf\,o}$, Fig. 6.7 (c). These support dimensions are rather small, considering the fact that

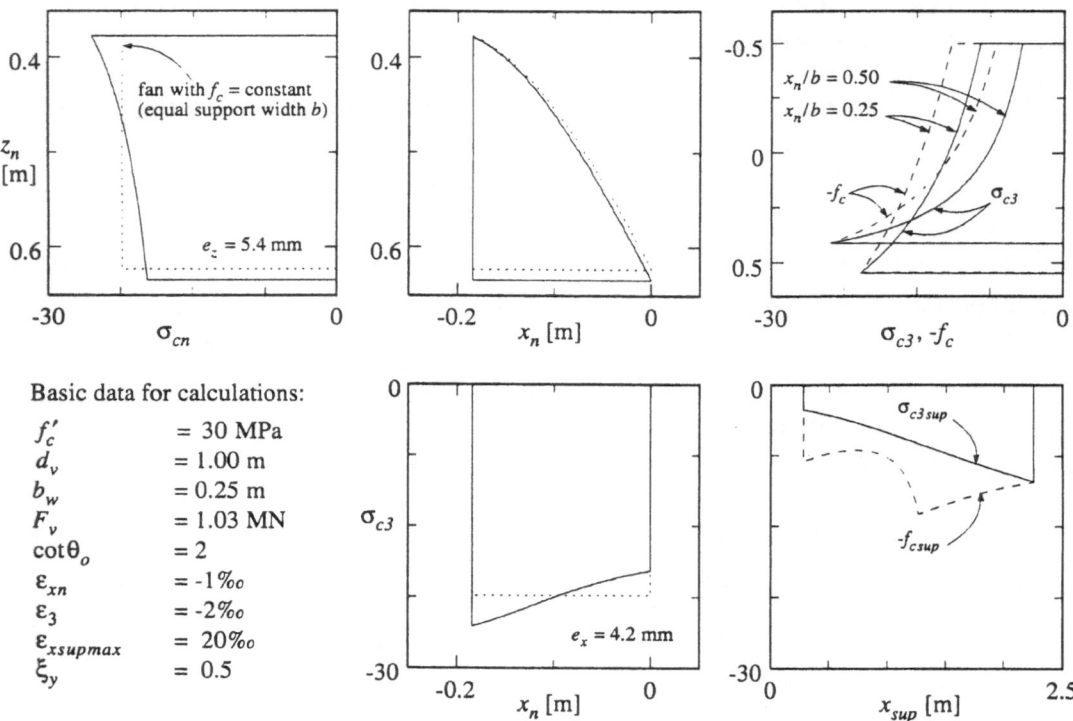

Fig. 6.8 – Discontinuity region, non-centred fan with variable f_c: results of example calculation. Note: stresses in MPa.

the concrete in the adjoining parallel stress band is at the onset of crushing and that a flat inclination θ_o has been chosen, $\cot\theta_o = 2$. The distribution of the stresses acting on the horizontal and vertical boundaries of the nodal zone is slightly non-uniform; however, the eccentricities e_x and e_z are small and could be neglected in design practice. An average concrete compressive strength of $f_c = 22.4$ MPa or $f_c = 2.32(f_c')^{2/3}$ in MPa had to be assigned to the non-centred fan with constant f_c in order to obtain the same support width b. According to the bottom right diagram, crushing of the web at the tension chord is not involved since the concrete stresses are below f_c along the entire tension chord and hence, concrete stresses do not exceed f_c throughout the fan region, Chapter 6.3.3. The top right diagram illustrates the hyperbolic variations of both f_c and the principal concrete compressive stresses σ_{c3} along two selected trajectories, cf. Fig. 6.7 (f).

Fig. 6.9 illustrates the influence of $\cot\theta_o$ on the support width b, on the depth of the nodal zone c, and on the average value of f_c to be assigned to a non-centred fan with f_c = constant in order to obtain the same support width b as from the calculations with variable f_c. As for Fig. 6.8, $(q + f_w) =$ constant has been assumed, reducing the support reaction to $F_v = (q + f_w)d_v\cot\theta_o = 0.7$ MN; this allowed investigating a wider range of $\cot\theta_o$. Whereas a constant web thickness of $b_w = 0.35$ m has been used in Fig. 6.9 (a), the value of b_w in Fig. 6.9 (b) has been chosen such that the concrete in the adjoining parallel stress band is at the onset of crushing for the given values of $\cot\theta_o$ and F_v,

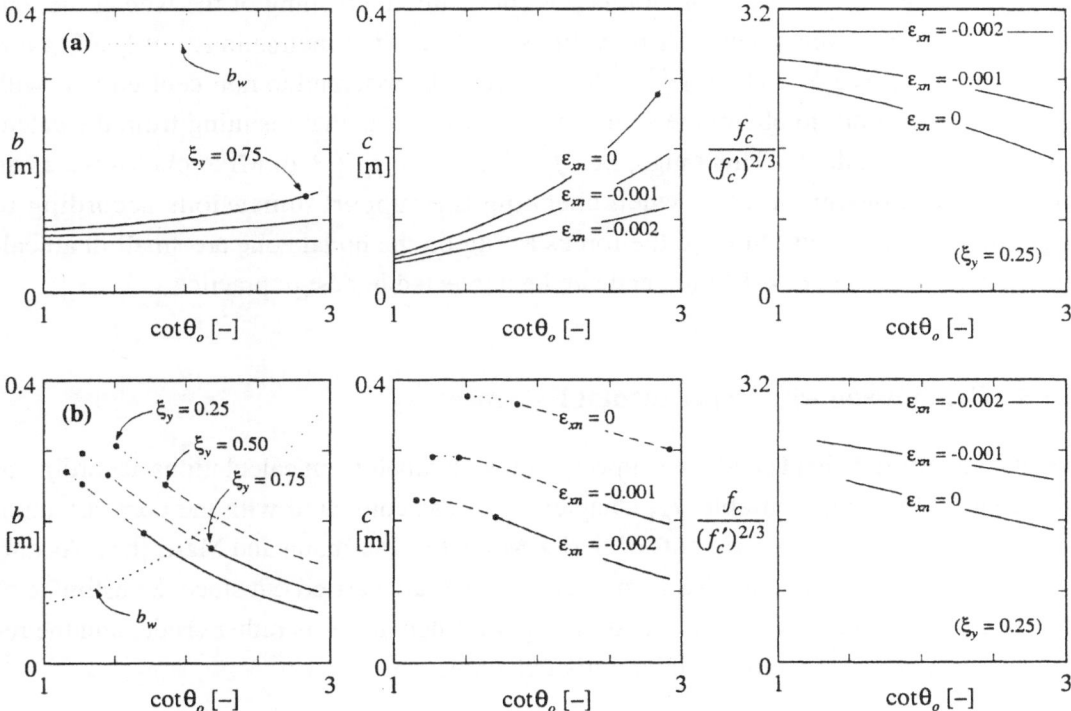

Fig. 6.9 – Non-centred fan: support width b, height of nodal zone c, and average concrete stresses over support for variable inclination θ_o and (a) b_w = constant; (b) b_w = minimum value for given θ_o. Note: f_c, f_c' in MPa.

$-\sigma_{c3} = f_c$ for $z = -d_v/2$ and $\xi = 1$. The calculations have been carried out for three different values of ε_{xn}, and the concrete stresses have been checked along the tension chord for three different values of ξ_y. Apart from the changes mentioned, the basic data indicated in Fig. 6.8 have been used for Fig. 6.9 as well.

From Fig. 6.9 (a) it can be seen that for b_w = constant, flat inclinations θ_o primarily affect the height of the nodal zone c since the horizontal component of the support reaction increases with $\cot\theta_o$. In reality, the horizontal stresses are partly carried by axial reinforcement and somewhat smaller values of c than indicated in Fig. 6.9 are possible. The influence of the inclination θ_o on the support width b is not very pronounced for b_w = constant since the steeper trajectories carry most of the applied load in either case, and crushing of the web at the tension chord is not critical apart from the rather unrealistic combination ($\xi = 0.75$, $\varepsilon_{xn} = 0$). Fig. 6.9 (b) illustrates that if the web thickness b_w is chosen such that the concrete in the adjoining parallel stress band is at the onset of crushing, wider supports are required for steeper inclinations θ_o. For flatter inclinations θ_o higher values of b_w are required in order to avoid web-crushing failures in the adjoining parallel stress band; thus, the steep trajectories of the fan can carry much more load. For the same reason, crushing of the web at the tension chord occurs for steep inclinations θ_o and high values of ξ_y. In both cases, illustrated in Figs. 6.9 (a) and (b), the axial strains ε_{xn} significantly influence the required support dimensions b and c as well as the values of $\cot\theta_o$ for which a solution is possible without crushing of the web at the tension chord. Apart from the unrealistic value $\varepsilon_{xn} = 0$ and flat inclinations $\cot\theta_o > 2$, average values of $f_c = 2.5 \ldots 3.0 (f_c')^{2/3}$ in MPa have to be assigned to non-centred fans with constant f_c in order to obtain the same support width b as that resulting from the calculations with variable f_c. This range, i.e. $f_c = 2.5 \ldots 3.0 (f_c')^{2/3}$ in MPa, may serve as an orientation in design practice when checking the support dimensions according to Eqs. (6.21). The eccentricities of the forces acting on the nodal zone are small in all calculations, $e_x < 8$ mm, $e_z < 17$ mm, and can be neglected in design practice.

6.3.5 Comparison with Experimental Evidence

Below, the ultimate loads and the support widths obtained from calculations according to non-centred fans with variable f_c, Chapter 6.3.4, are compared with the corresponding values observed in the experiments of Series MVN by Kaufmann and Marti [65]. As outlined in Chapter 6.3.1, no load-deformation analyses are carried out since the estimate of the state of strain in the fan region underlying the calculations is rather crude, and the results of such analyses would be inherently debatable.

Fig. 6.10 shows the test setup of the experiments of Series MVN; like Series VN, Fig. 6.5, the experiments were conducted in the "Beam Element Tester" [65] of the Swiss Federal Institute of Technology, Zurich. Specimen geometry and transverse reinforcement were the same as in Series VN (length 5.84 m, height 0.78 m, doubly symmetrical cross-section, web thickness $b_w = 0.15$ m, flange width 0.80 m, flange height 0.14 m, closed stirrups $\varnothing = 8$ mm spaced at 0.20 m). Series MVN consisted of two conventionally reinforced and two prestressed girders and corresponded to regions of girders near intermediate supports or concentrated loads. The test parameters were the applied

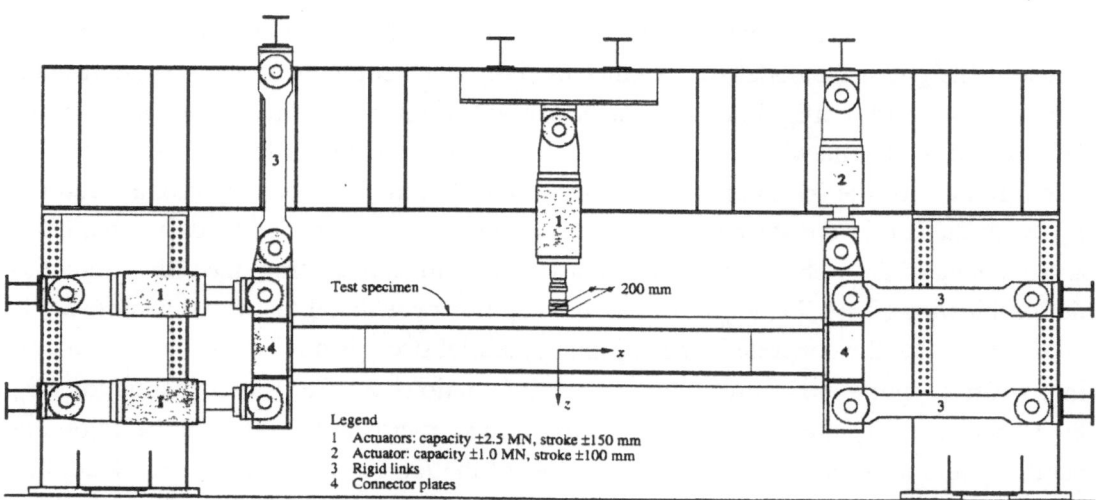

Fig. 6.10 – Experiments by Kaufmann and Marti [65]: Elevation of the Beam Element Tester, test setup for Series MVN; scale 1:80.

(a)

Girder	MVN1	MVN2	MVN3	MVN4	Units
Specimen data					
N	–	-1.30	–	-1.30	[MN]
P_o	–	–	1.30	1.30	[MN]
f'_c	62.8	58.8	62.2	64.3	[MPa]
ε_{co}	2.23	2.25	2.40	2.49	[‰]
Conditions at the ultimate state					
ε_{xexp} [a]	-0.80	-0.91	-1.06	-1.78	[‰]
	-2.48	-2.41	-2.81	-3.30	
F_{vuexp}	.532	.604	.680	.775	[MN]

[a] upper values: flange centre (interpolated); lower values: top of girder.

(b)

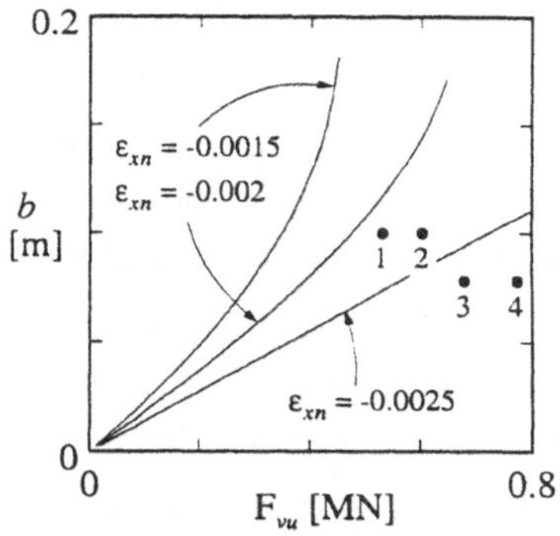

Data used in calculations:

f'_c = 62 MPa
ε_{co} = 2.5‰
d_v = 0.64 m
$\cot\theta_o$ = 3.97

Fig. 6.11 – Non-centred fan, concrete strength according to Eq. (5.4): (a) basic data and summary of results of experiments of Series MVN by Kaufmann and Marti [65]; (b) calculated and observed conditions at the ultimate state.

axial force N and the initial prestressing force P_o, see Fig. 6.11 (a). The grouted post-tensioning tendons in the prestressed specimens MVN3 and MVN4 were inclined at $\tan\alpha_p = 0.1115$ from the beam centres at the element ends to midspan where they were deviated parabolically on a length of 1 m (radius of curvature 4.5 m). Loads were applied at the element ends as well as at midspan (support plate 0.20·0.25 m, Fig. 6.10). In all experiments, axial forces (if any) were applied first and subsequently the force at midspan was gradually increased up to failure. The forces applied at the element ends were controlled such that axial forces remained constant throughout the test and shear forces on either side of the concentrated load at midspan were of equal magnitude. In a first phase forces were controlled such that no moments developed at the element ends. After the yield moment at midspan had been reached, rotations of the connector plates, Fig. 6.10, were fixed, and moments developed at the element ends. This second phase of the tests simulated a redistribution of moments from the support region (midspan) into the span (element ends) in a continuous girder. In all four tests collapse was governed by failure of the web in the central portion of the girders. Failure of Specimen MVN1 was triggered by the rupture of a stirrup after some spalling of the web cover concrete. Specimens MVN2 and MVN3 failed in a very brittle manner; half of the girder was shifted over the remaining part of the specimen along an inclined failure surface. Specimen MVN4 exhibited a more ductile failure; the web crushed progressively on both sides of the girder.

Fig. 6.11 (b) compares the support widths and the ultimate loads observed in the experiments of Series MVN with calculations using a non-centred fan with f_c according to Eq. (5.4). Since the specimens were symmetrical about midspan, only half of the element has been considered, assigning half of the force $2F_v$ acting at midspan to either half-specimen. Thus, the experimental values of the ultimate loads F_{vu} indicated in Fig. 6.11

for the non-prestressed experiments MVN1 and MVN2 correspond to half of the experimentally observed loads at midspan, and the experimental support widths are equal to half of the width of the support plate, $b = 0.10$ m. The deviation forces $2F_d$ of the post-tensioning tendon, acting at midspan of the prestressed specimens MVN3 and MVN4, have been accounted for as concentrated forces, reducing the load carried by the fans by $2F_d$ but using up the central portion of $2F_d/(b_w f_c')$ of the support plate. Assuming that the prestressing steel yielded at the ultimate state, $P_y = 1802$ kN [65], one obtains $F_d \approx P_y \tan \alpha_p = 201$ kN per half-specimen. Consequently, the values of F_{vu} and b indicated in Fig. 6.11 for Specimens MVN3 and MVN4 have been obtained by deducting $F_d = 201$ kN and $F_d/(b_w f_c') \approx 22$ mm from half of the experimentally observed loads at midspan and half of the width of the support plate, respectively.

The calculations have been carried out for an effective shear depth of $d_v = 0.64$ m, corresponding to the distance between the centres of the flanges. A uniform distribution of the stirrup forces $f_w = F_v/(d_v \cot\theta_o)$ has been assumed ($q=0$), selecting the inclination $\cot\theta_o = 3.97$ such that the flattest trajectory of the fan, starting at the edge of the support plate at midspan, intersects the lower chord ($z=d_v/2$) at the element ends. Although the transverse reinforcement was stronger at the element ends, the assumption of constant stirrup forces appears to be reasonable in the light of the very flat inclinations of the corresponding trajectories of the fan. In the central portion of the elements, uniformly distributed loads $f_w = F_v/(d_v \cot\theta_o)$ would have corresponded to stirrup stresses of 417, 473, 532 and 607 MPa for Specimens MVN1 to MVN4, respectively; these values should be compared with the yield and ultimate strengths of the stirrups, $f_{syz} = 461$ MPa and $f_{suz} = 580$ MPa [65]. Thus, the assumed uniform distribution of the stirrup forces can be justified for Specimens MVN1 to MVN3, whereas in Specimen MVN4, either the flattest trajectories carried somewhat higher stresses or the deviation force of the post-tensioning tendon was slightly higher than according to the assumptions made. Since the concrete properties were similar in all four specimens, Fig. 6.11 (a), average values of $f_c' = 62$ MPa and $\varepsilon_{co} = -\varepsilon_3 = 2.5\%$ have been applied in all calculations, and rather than using the values of the axial strains of the compression chord observed in the tests, calculations have been carried out for three different values of ε_{xn}. This allows plotting all four experiments in the same diagram, Fig. 6.11 (b). Failure caused by crushing of the concrete at the tension chord has been excluded from the present calculations.

Fig. 6.11 (b) demonstrates that the axial strains ε_{xn} along the boundary of the nodal zone significantly influence the support width b required for a given support reaction F_v. Considering the experimentally observed values of ε_x indicated in Fig. 6.11 (a) and noting that these strains were recorded on the surface of the flange rather than at the actual boundary of the nodal zone, the agreement of predicted and observed ultimate loads for the given support widths is satisfactory. It should be emphasised that both the experimental values and the predicted curves indicated in Fig. 6.11 (b) involve several assumptions and that the present comparison of ultimate loads and support widths, though satisfactory, can only be regarded as a verification of the order of magnitude of results obtained from calculations using non-centred fans with f_c according to Eq. (5.4).

6.4 Additional Considerations

6.4.1 General Remarks

The general procedure and the approximate solution for low gradients of the stirrup stresses presented in Chapter 6.2 allow performing load-deformation analyses of the web of concrete girders in continuity regions, accounting for tension stiffening effects of the stirrups and variations of the principal concrete compressive stress direction θ in the web. Per se such analyses are of little practical importance since shear deformations contribute comparatively little to the overall deformations of girders [144]. However, they can be used to investigate the appropriateness of the typical design assumption of a uniform uniaxial compressive stress field in the web, see Chapter 4.2. Nearly straight principal compressive stress trajectories in the web are indeed obtained at the ultimate state as demonstrated by the numerical example, Fig. 6.4, even for small stirrup reinforcement ratios causing substantial redistribution of the internal forces. Thus, the application of parallel stress bands to the design of continuity regions is justified by the present work.

Premature failure by crushing of the concrete can be prevented by comparing the principal concrete compressive stresses in the web with the effective concrete compressive strength f_c according to Eq. (5.4), or by directly comparing the applied shear stresses with the shear resistance governed by crushing of the concrete, Eq. (5.32)$_2$. As proposed by the Canadian Code [20], Chapter 4.3.2, one may assume $\varepsilon_3 = -\varepsilon_{co}$, and the axial strains ε_x can be determined from a strain-compatibility analysis for the section under the action of the moment M and the axial force $N + V\cot\theta$, where N and V are the applied axial and shear forces, Fig. 4.3. Basically, the value of ε_x used in assessing f_c should be determined at the tension chord where the principal tensile strains ε_1 are highest. However, the webs of profiled girders are laterally restrained by the chords, resulting in an enhancement of the concrete compressive strength as confirmed by the experiments of Series VN, Chapter 6.2.5. Hence, ε_x can be determined some distance away from the tension chord, e.g. at a distance b_w from the inner edge of the flange, and in girders of moderate height, the value of ε_x at mid-depth of the web can be used.

Care should be taken with respect to the ductility of the stirrups, particularly for flat inclinations θ. In three tests of Series VN, Chapter 6.2.5, failure was triggered by stirrup rupture although the ductility properties of the reinforcement were at the limit between the ductility classes N and H (normal and high ductility) according to Eurocode 2 [48]. The approximate solution presented in Chapter 6.2.4 can be used to check the ductility of the stirrups in continuity regions. Alternatively, the vertical strains at the tension chord, $\varepsilon_z = \varepsilon_3 + (\varepsilon_x - \varepsilon_3)\cot^2\theta$, can be compared with the rupture strain ε_{su} of the naked stirrups, noting that the overall strains at stirrup rupture may be much lower than ε_{su}, depending on the bond behaviour and the ductility properties of the stirrups, Chapter 2.4.2.

The fan-shaped discontinuous stress fields with variable f_c according to Eq. (5.4) presented in Chapter 6.3 allow verifying the design procedure for support regions suggested in [143,144]. According to this procedure, the web thickness b_w is determined from the adjoining parallel stress band as outlined above for continuity regions, whereas the support dimensions b and c as well as the assumed values of d and d_v are checked separately, using Eqs. (6.21) and a comparatively high value of f_c. As demonstrated by the calculations underlying Fig. 6.9 the concrete stresses along the boundary of the nodal zone of the fan are below f_c according to Eq. (5.4) if the support dimensions are determined from Eqs. (6.21), using an average value of $f_c = 2.5...3.0(f_c')^{2/3}$ in MPa. On the other hand, according to Chapter 6.3.3, checking the concrete stresses at the top and bottom end of the fan is sufficient to ensure that the concrete stresses are below f_c throughout the fan region. Hence, the suggested design procedure for support regions [143,144] can be justified by the present work if the principal concrete compressive stresses are checked along the tension chord and a value of $f_c = 2.5...3.0(f_c')^{2/3}$ in MPa is used to determine the support dimensions from Eqs. (6.21).

6.4.2 Prestressing and Axial Forces

Straight prestressing tendons and axial forces in girders can be accounted for similarly as in membrane elements, Chapter 5.4.2. The remarks made in Chapter 5.4.2 on the influence of axial forces on the load-carrying behaviour apply correspondingly, noting that longitudinal forces are assigned to the chords by the models presented in Chapters 6.2 and 6.3.

Straight or curved prestressing tendons can be included in the models presented in Chapters 6.2 and 6.3 by accounting for anchor, deviation and friction forces as applied loads [107,146]. Typically, prestressing tendons are initially stressed to about $0.7f_{pu}$, corresponding to roughly $(0.8...0.85)f_{py}$; when determining ultimate loads, the full yield strength f_{py} is usually applied at the locations where the peak moments occur. The required increase of the force in the prestressing tendon can be modelled by suitable discontinuous stress fields [66,146,147], similar to that illustrated in Fig. 6.7 (b). However, such stress fields are difficult to establish, particularly if the tendon forces increase in regions where the tendon is curved [147]. Hence, increases of the prestressing forces have often been neglected in stress field solutions.

6.4.3 Girders without Shear Reinforcement

The models considering rotating, stress-free cracks presented in Chapters 6.2 and 6.3 predict zero shear resistance for girders without shear reinforcement, $\rho_z = 0$, unless compressive stresses $\sigma_z < 0$ are applied. As in uniaxially reinforced membrane elements, Chapter 5.4.3, shear stresses τ_{xz} can only be resisted by girders without shear reinforcement and $\sigma_z \geq 0$ if tensile stresses are allowed in the web. In continuity regions, load-carrying mechanisms as illustrated in Fig. 5.13 are possible if the longitudinal reinforcement is strong, and thus, girders without shear reinforcement will not necessarily fail upon initial cracking of the web; further details are given in Chapter 5.4.3.

7 Summary and Conclusions

7.1 Summary

This thesis aims at contributing to a better understanding of the load-carrying and deformational behaviour of structural concrete subjected to in-plane shear and normal forces, including membrane elements (homogeneous state of plane stress) and webs of girders with profiled cross-section (non-homogeneous state of plane stress). Simple, consistent physical models reflecting the influences of the governing parameters are developed on whose basis (i) a realistic assessment of the deformation capacity of structural concrete subjected to in-plane loading is possible, (ii) the limits of applicability of the theory of plasticity to structural concrete can be explored, and (iii) current design provisions can be critically reviewed, supplemented and harmonised.

The first part of this thesis (Chapters 2 to 4) covers material properties, basic aspects of the theory of plasticity, and a survey of previous work on plane stress problems.

In Chapter 2 the properties of concrete and reinforcement relevant for structural concrete subjected to in-plane stresses are examined and existing models for the behaviour of concrete are compared with the results of recent tests on high-strength concrete specimens. Based on the consideration of different failure modes a possible explanation for the experimentally observed fact that the uniaxial compressive strength of laterally unconstrained concrete specimens increases less than proportional with respect to the cylinder strength is presented. The behaviour of plain concrete subjected to biaxial loading as well as triaxial compressive stresses is reviewed, and the basic mechanisms of aggregate interlock behaviour are summarised. Fundamental aspects of bond behaviour are outlined and tension stiffening effects are examined using a bilinear stress-strain characteristic for the reinforcement and a stepped, rigid-perfectly plastic bond shear stress-slip relationship as proposed by Sigrist [144]. This idealisation is referred to as tension chord model [144,7,93,6,94] and is used to account for tension stiffening effects throughout Parts 2 and 3 of the present thesis. The behaviour of structural concrete subjected to biaxial compression and tension is examined and a relationship for the compression softening of the concrete is presented, Eq. (2.27). The proposed relationship accounts for the deteriorating influence of lateral tensile strains on the concrete compressive strength as well as for the observation that the uniaxial compressive strength of laterally unconstrained concrete specimens increases less than proportional with respect to the cylinder strength. The relationship is shown to correlate well with the experimental evidence and is used throughout Parts 2 and 3 of the present thesis to account for compression softening effects of the concrete.

In Chapter 3 the theory of plastic potential and the basic theorems of limit analysis for perfectly plastic materials are summarised and basic aspects of the application of limit analysis methods to structural concrete are examined.

In Chapter 4 previous work on plane stress problems is reviewed. After a brief historical account limit analysis approaches for membrane elements and beams are examined in Chapter 4.2, and fundamental aspects of the behaviour of cracked concrete membranes and compression field approaches are investigated in Chapter 4.3. A general numerical procedure [67] that allows one to treat cracks as fixed and interlocked rather than as rotating and stress-free is outlined, and it is emphasised that this procedure is the most general possible approach provided that only one set of cracks and uniform distributions of steel and bond shear stresses in the transverse direction between the individual reinforcing bars are considered. Based on a subdivision of total strains into concrete strains and strains due to cracks a simplified approach that allows one to treat cracks as fixed and interlocked rather than as rotating and stress-free is also presented. Finally, previous compression field approaches are critically reviewed, clarifying underlying assumptions and pointing out conceptual peculiarities, and the relationships between the different approaches are discussed.

The second part of this thesis, Chapter 5, covers the behaviour of membrane elements subjected to a homogeneous state of plane stress.

In Chapter 5.2 a new model for cracked, orthogonally reinforced concrete panels subjected to a homogeneous state of plane stress is presented. The cracked membrane model combines the basic concepts of the original compression field approaches and the tension chord model. Crack spacings and tensile stresses between the cracks are determined from first principles and the link to limit analysis methods is maintained since equilibrium conditions are expressed in terms of stresses at the cracks rather than average stresses between the cracks as in previous, similar approaches. Both a general numerical method suitable for performing comprehensive load-deformation analyses and an approximate analytical solution are derived, including expressions for crack spacings and widths. In Chapter 5.3, the results are compared with previous theoretical and experimental work. It is shown that the approximate analytical solution comprises the solution derived by Baumann [10] as a special case, and that the diagonal crack spacing resulting from the underlying derivation corresponds to the expression suggested by Vecchio and Collins [156] without further justification. Very good agreement between the experimentally observed load-deformation behaviour and the calculations according to the general numerical method of the cracked membrane model is obtained. Furthermore, based on a detailed comparison with limit analysis methods, simple expressions for the ultimate load of reinforced concrete panels in terms of the reinforcement ratios and the cylinder compressive strength of concrete are presented, Eq. (5.33), and the expressions are shown to correlate well with the experimental evidence. In Chapter 5.4 the influences of prestressing and axial forces are examined and basic aspects of the load-carrying behaviour of uniaxially reinforced membrane elements are discussed.

The proposed compression softening relationship and the simple expressions for the ultimate load of reinforced concrete panels established in Chapter 5 are shown to correlate well with experimental results obtained from numerous tests on orthogonally reinforced membrane elements subjected to in-plane shear and normal forces which failed by crushing of the concrete while one or both reinforcements remained elastic. Experimental data and underlying calculations are summarised in Appendix B.

In the third part of this thesis, Chapter 6, the behaviour of beams in shear is examined, focusing on simplified models for girders with flanged cross-section.

In Chapter 6.2 situations where all static and geometric quantities vary only gradually along the girder axis are investigated, assuming that the state of stress in the web is constant along the girder axis. Both a general procedure as well as an approximate solution for low gradients of the stirrup stresses are presented. These procedures allow determining the states of stress and strain in the web and they enable one to carry out comprehensive load-deformation analyses, accounting for finite crack spacings, tension stiffening of the stirrups, compression softening of the concrete as well as for the variation of the principal compressive stress (and strain) direction over the depth of the cross-section. Furthermore, an analytical solution for cases in which bond slip occurs in the same direction over the entire depth of the cross-section is given. The results are compared with typical design assumptions and with previous experimental work. Good agreement between the experimentally observed load-deformation behaviour and the calculations according to the approximate solution for low gradients of the stirrup stresses is obtained. Furthermore, it is demonstrated that even for small stirrup reinforcement ratios necessitating substantial redistribution of the internal forces nearly straight principal compressive stress trajectories in the web are obtained at the ultimate state. Thus, the typical design assumption of a uniform uniaxial compressive stress field in the web is justified. Simplified approaches that allow preventing premature failure by crushing of the concrete in the web or by stirrup rupture are also proposed.

In Chapter 6.3 discontinuity regions characterised by abrupt changes of static quantities are analysed using the practically relevant case of the support region of a constant-depth girder with flanged cross-section as an illustrative example. Fan-shaped discontinuous stress fields for variable concrete compressive strength are examined, and a method that allows checking whether the concrete stresses are below the concrete compressive strength – accounting for compression softening of the concrete – throughout the fan region is presented. It is pointed out that according to the assumptions made checking the concrete stresses at the top and bottom end of the fan-shaped stress field is sufficient to ensure that the concrete stresses are below the concrete compressive strength throughout the fan region. Some numerical examples are presented and the results are compared with typical design assumptions and with previous experimental work. Satisfactory agreement of predicted and observed ultimate loads for given support widths is obtained, and a previously suggested design procedure for support regions [143,144] is supplemented and justified.

7.2 Conclusions

Due to the treatment of tension stiffening effects according to the tension chord model and the adoption of a compression softening relationship reflecting the governing influences, the cracked membrane model presented in Chapter 5 generally results in excellent predictions of the load-deformation behaviour, including cases where failure is governed by crushing of the concrete. On this basis a realistic assessment of the deformation capacity of structural concrete elements subjected to in-plane loading is possible and current design provisions can be critically reviewed, supplemented and harmonised.

In design practice, the different methods presented in Chapters 5.2 and 5.3 should be applied judiciously, depending on the nature of the problem under consideration and the required level of sophistication. Frequently, the determination of failure conditions according to limit analysis is sufficient; comprehensive load-deformation analyses are only rarely required. Ultimate loads can be accurately predicted by Eq. (5.33), whereas the approximate analytical solution, Eq. (5.17), may be used for a simplified analysis of the cracked elastic response. The load-deformation behaviour according to the general numerical method of the cracked membrane model is well approximated by the combined application of Eqs. (5.17) and (5.33).

The general procedure and the approximate solution for low gradients of the stirrup stresses presented in Chapter 6.2 allow performing load-deformation analyses of the web of concrete girders in continuity regions, accounting for the variation of the principal compressive stress direction over the depth of the cross-section. As for membrane elements the present treatment of tension stiffening and compression softening effects results in good predictions of the load-deformation behaviour. While such analyses are of little practical importance (since shear deformations contribute little to the overall deformations of girders), they can be used to investigate the appropriateness of the typical design assumption of a uniform uniaxial compressive stress field in the web. As demonstrated in Chapter 6.2 nearly straight principal compressive stress trajectories in the web are indeed obtained at the ultimate state, even for small stirrup reinforcement ratios necessitating substantial redistribution of the internal forces. Hence, the parallel stress bands and the sectional design approach presented in Chapter 4.2 are justified, provided that premature failures by crushing of the concrete or stirrup rupture are prevented by checking concrete stresses and stirrup strains as proposed in Chapter 6.4.1.

The design procedure for support regions suggested in [143,144] has been justified and supplemented by the fan-shaped discontinuous stress fields with variable concrete compressive strength presented in Chapter 6.3. Safe designs are obtained if (i) the web thickness is determined such that web crushing failure in the adjoining parallel stress band is prevented, (ii) the support dimensions are determined from Eq. (6.21) using a value of $f_c = 2.5...3.0(f_c')^{2/3}$ in MPa at the support, and (iii) the principal concrete compressive stresses are checked along the tension chord, accounting for the plastic chord strains caused by a possible moment redistribution.

7.3   Recommendations for Future Research

The cracked membrane model, Chapter 5, and the procedures that allow to treat cracks as fixed and interlocked rather than as rotating and stress-free, Chapter 4, are very promising and may serve as the starting point for future research along the following lines:

- The extension of the tension chord model to cracked panels could be completed by comparing the expressions for crack spacings and crack widths presented in this work with the experimental evidence. This would allow extending the range of application of the cracked membrane model to serviceability and crack width calculations.

- The procedures that allow one to treat the cracks as fixed and interlocked – which include the cracked membrane model considering rotating, stress-free cracks as a special case – could be implemented in a computer program suitable for deformation capacity considerations, making use of a plane stress element formulation in an existing non-linear finite element code. This would allow comparing overall response predictions according to the different approaches and hence evaluating the influence of considering the cracks as fixed and interlocked rather than as rotating and stress-free.

- Starting from the cracked membrane model the demand for and the supply of deformation capacity of entire wall structures could be investigated, giving special consideration to possible restraint actions and to the necessary amounts and distributions of minimum reinforcement required to avoid premature failure and excessive crack widths under service conditions. The demand for deformation capacity cannot be determined based purely on an element consideration since any redistribution of internal forces involves the entire structure. Hence, refined models – presumably based on discontinuous stress field analyses similar to the approach applied in [144] – would have to be applied, and finite element analyses based on the cracked membrane model, using the plane stress element formulation mentioned above, would be helpful.

- Adopting a sandwich model approach [88,89] the basic concepts of the cracked membrane model could be extended to plate structures (walls, slabs and shells) which are generally subjected to combined moments, membrane forces and transverse shear forces. Due to the new treatment of tension stiffening effects in the top and bottom layers according to the cracked membrane model such a sandwich model would allow one to assess the influence of the bond characteristics and the ductility of the reinforcement on the deformation capacity of plate structures and to study compression softening effects of the concrete in over-reinforced plate elements which fail by crushing of the concrete while the reinforcement remains elastic. The information gathered in this way could be used to assess the amount of redistribution of internal forces in plate structures permissible in design.

Furthermore, discontinuous stress fields similar to those presented in Chapter 6.3, but capable of modelling force increases in curved prestressing tendons, could be analysed in order to explore the limits of applicability of the typical design assumption that at the ultimate state, the prestressing steel yields at locations where the peak moments occur.

Appendix A: Characteristic Directions in Plane Stress

Following similar lines as in [82,109], the governing quasi-linear partial differential equations for the ultimate state of stress in a perfectly plastic material following the associated flow rule will be established, and the characteristic directions in plane stress will be derived. Since, in plane stress, the modified Coulomb failure criterion can be represented by two Mohr failure envelopes [82], corresponding expressions follow as special cases from the more general analysis for plane strain and a yield condition of the Mohr type, $|\tau| - g(\sigma) \leq 0$.

As pointed out by Marti [82], any failure criterion of an isotropic, perfectly plastic material in plane strain can be expressed by a condition of the type

$$q = q(p) \tag{A.1}$$

where $p = (\sigma_1 + \sigma_3)/2$ and $q = (\sigma_1 - \sigma_3)/2$ are the abscissa of the centre and the radius of the governing Mohr's circle of stress, as illustrated in Fig. A.1 (a). The cited reference also discusses the conditions for which a failure criterion of the type (A.1) can be reduced to a yield condition of the Mohr type, $|\tau| - g(\sigma) \leq 0$. As already mentioned above, this is possible for the modified Coulomb failure criterion in plane stress and hence, Eq. (A.1) can be rewritten as

$$\frac{dq}{dp} = -\sin\varphi(p) \tag{A.2}$$

see Fig. A.1 (a).

Equilibrium of an infinitesimal element of unit thickness subjected to body forces f_x and f_z, Fig. A.1 (b), requires that

$$\begin{aligned} \sigma_{x,x} + \tau_{xz,z} + f_x = 0 \\ \tau_{zx,x} + \sigma_{z,z} + f_z = 0 \end{aligned} \tag{A.3}$$

where the comma designates partial derivatives with respect to the variable following the comma.

The stress components σ_x, σ_z, and τ_{xz} can be expressed in terms of p, q and the inclination θ of the principal compressive stress direction with respect to the x-axis, see Fig. A.1 (c), i.e.

$$\begin{aligned} \sigma_x &= p - q\cos(2\theta) \\ \sigma_z &= p + q\cos(2\theta) \\ \tau_{xz} &= -q\sin(2\theta) \end{aligned} \tag{A.4}$$

Thus, using Eqs. (A.1) and (A.2), one gets

$$\sigma_{x,x} = [1 + \sin\varphi\cos(2\theta)]p_{,x} + 2q\sin(2\theta)\theta_{,x}$$

$$\sigma_{z,z} = [1 - \sin\varphi\cos(2\theta)]p_{,z} - 2q\sin(2\theta)\theta_{,z}$$

$$\tau_{zx,x} = \sin\varphi\sin(2\theta)p_{,x} - 2q\cos(2\theta)\theta_{,x}$$ (A.5)

$$\tau_{xz,z} = \sin\varphi\sin(2\theta)p_{,z} - 2q\cos(2\theta)\theta_{,z}$$

Substituting these expressions in Eq. (A.3) and rearranging, one obtains a system of two quasi-linear partial differential equations for the unknown functions $p = p(x,z)$ and $\theta = \theta(x,z)$

$$\mathbf{C} \cdot \begin{bmatrix} p_{,x} \\ p_{,z} \\ \theta_{,x} \\ \theta_{,z} \end{bmatrix} + \begin{bmatrix} f_x \\ f_z \end{bmatrix} = \begin{bmatrix} 0 \\ 0 \end{bmatrix}$$ (A.6)

where $\mathbf{C}$ is given by

$$\mathbf{C} = \begin{bmatrix} 1 + \sin\varphi\cos(2\theta) & \sin\varphi\sin(2\theta) & 2q\sin(2\theta) & -2q\cos(2\theta) \\ \sin\varphi\sin(2\theta) & 1 - \sin\varphi\cos(2\theta) & -2q\cos(2\theta) & -2q\sin(2\theta) \end{bmatrix}$$ (A.7)

Assuming that $p(x,z)$ and $\theta(x,z)$ are a solution of Eq. (A.6), a one-dimensional manifold within that solution can be defined by an equation of the type $\psi(x,z) = 0$; within

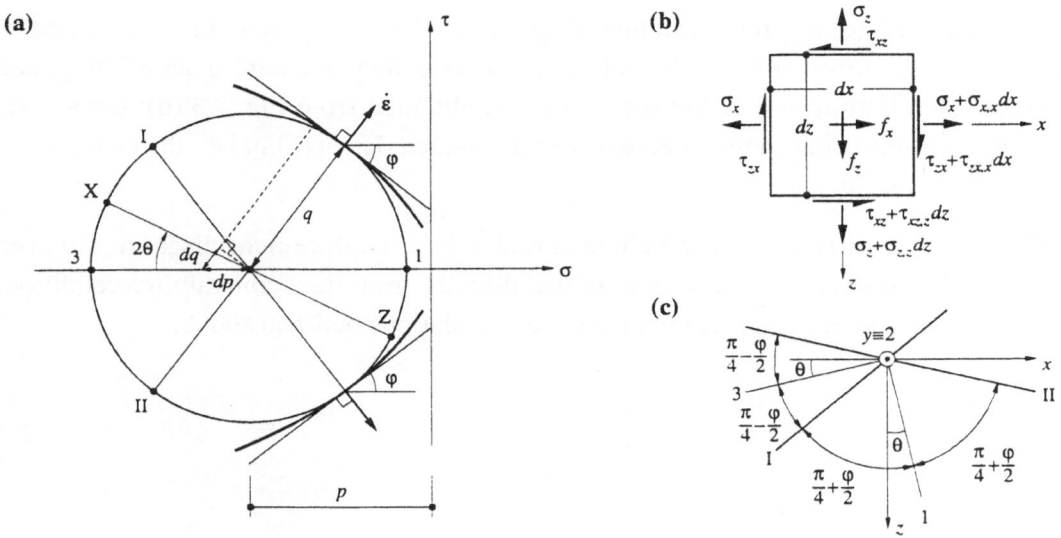

Fig. A.1 – Directions of characteristic curves: (a) yield condition and governing stress circle; (b) infinitesimal element; (c) principal directions and directions of the characteristics.

129

$\psi(x,z)$, a parameter λ is introduced. Considering $\psi(x,z) = 0$ and λ rather than x and z as the independent variables [37], Eq. (A.6) transforms to

$$\begin{bmatrix} c_{11}\Psi_{,x} + c_{12}\Psi_{,z} & c_{13}\Psi_{,x} + c_{14}\Psi_{,z} \\ c_{21}\Psi_{,x} + c_{22}\Psi_{,z} & c_{23}\Psi_{,x} + c_{24}\Psi_{,z} \end{bmatrix} \cdot \begin{bmatrix} p_{,\psi} \\ \theta_{,\psi} \end{bmatrix} + \begin{bmatrix} g_1 \\ g_2 \end{bmatrix} = \begin{bmatrix} 0 \\ 0 \end{bmatrix} \tag{A.8}$$

where c_{ij} are the components of matrix $\mathbf{C}$, Eq. (A.7), and g_1 and g_2 designate arbitrary functions which do not depend on the partial derivatives leading out of the one-dimensional manifold $\psi(x,z) = 0$, i.e., g_1 and g_2 are independent of $p_{,\psi}$ and $\theta_{,\psi}$.

From Eq. (A.8) it can be seen that $p_{,\psi}$ and $\theta_{,\psi}$ and hence, all other partial derivatives leading out of the one-dimensional manifold ψ are uniquely determined if the values of p and θ are given along the curves $\psi(x,z) = 0$ in the (x,z)-plane, unless

$$\left\| \begin{bmatrix} c_{11}\Psi_{,x} + c_{12}\Psi_{,z} & c_{13}\Psi_{,x} + c_{14}\Psi_{,z} \\ c_{21}\Psi_{,x} + c_{22}\Psi_{,z} & c_{23}\Psi_{,x} + c_{24}\Psi_{,z} \end{bmatrix} \right\| = 0 \tag{A.9}$$

The solution $p(x,z)$, $\theta(x,z)$ of Eq. (A.6) may thus be discontinuous across the curves defined by $\psi(x,z) = 0$ if and only if the so-called characteristic condition, Eq. (A.9), is satisfied. Evaluating Eq. (A.9) and rearranging, one obtains the condition

$$\frac{dz}{dx} = \cot\left[\theta \pm \left(\frac{\varphi}{2} + \frac{\pi}{4} \right) \right] \tag{A.10}$$

The curves determined by Eq. (A.10) are the so-called characteristics (I,II) of the system of quasi-linear partial differential equations (A.6).

From Fig. A.1 (c) it can be seen that the principal directions 1 and 3 bisect the angles between the directions of the characteristics as in a displacement discontinuity, see Chapter 3.2.4. Setting $\alpha = \varphi$, the strains can be obtained from Fig. 3.3 (b); obviously, pure shear strains occur in the directions of the characteristics (I,II), i.e., the characteristics are inextensible.

Knowing the directions of the characteristics in a displacement discontinuity, see Chapter 3.2.4, Eq. (A.10) can be determined directly from the compatibility condition, Chapters 3.2.2 and 3.2.4, rather than from the calculations outlined above.

Appendix B:
Calibration of Proposed Compression Softening Relationship

In Chapter 2.4.3 the compression softening behaviour of concrete, i.e., the degradation of the concrete compressive strength by lateral tensile strains, has been examined and the expression

$$f_c = \frac{(f_c')^{2/3}}{0.4 + 30\varepsilon_1} \leq f_c' \qquad \text{in MPa} \tag{2.27}$$

for the concrete compressive strength has been proposed. Eq. (2.27) accounts for the influence of f_c' as well as that of ε_1 on f_c and correlates well with the results of tension-compression tests with reinforcement in the tensile direction, see Fig. 2.15.

The state of strain in tension-compression tests with reinforcement in the tensile direction differs from that in orthogonally reinforced elements subjected to in-plane shear forces, see Chapter 2.4.3. Nevertheless, good correlation with the experimental evidence has been obtained from the application of Eq. (2.27) to the compression softening of the concrete in cracked membrane elements and in the webs of girders with flanged cross-section, see Chapters 5.3.3, 6.2.5 and 6.3.5. Below, Eq. (2.27) is further validated by a comparison with the results of tests on orthogonally reinforced membrane elements. For comparison purposes the expression $f_c = f_c'(0.8 + 170\varepsilon_1)^{-1} \leq f_c'$ proposed by Vecchio and Collins [154,155], Eq. (2.25), and the simple relationship $f_c = 0.5\,f_c'$ for the concrete compressive strength are examined as well.

Tab. B.1 summarises basic data and results of experiments of Series PV by Vecchio and Collins [155], Series PHS by Vecchio, Collins and Aspiotis [158], Series TP and KP by André [9], Series SM by Ohmori et al. [116], Series SE by Khalifa [68], Kirschner and Collins [70], and Collins and Porasz [31], Series PP by Marti and Meyboom [91], Series A and B by Pang and Hsu [117], Series VA and VB by Zhang and Hsu [169,170], Series S by Yamaguchi and Naganuma [167], Series SP by Sumi and Iida [148], Series A, B and C by Sumi [149], Series A, B and C by Matuura and Sumi [100], and Series PL by Watanabe and Muguruma [163]. Square, orthogonally reinforced panels of constant thickness were subjected to monotonically increasing in-plane shear and normal forces in all experiments except Series SP [148]. In these tests, cylindrical shells were subjected to constant axial stresses while the applied shear stresses (torque) were monotonically increased until failure after 5 or 6 cycles of reversed shear (cyclic torques much lower than the ultimate load). Tab. B.1 includes 45 isotropically reinforced specimens, $\tilde{\rho}_x = \tilde{\rho}_z$, and 33 orthotropically reinforced specimens, $\tilde{\rho}_x > \tilde{\rho}_z$, where $\tilde{\rho}_x = \rho_x - \sigma_x/f_{syx}$ and $\tilde{\rho}_z = \rho_z - \sigma_z/f_{syz}$ are the effective reinforcement ratios in shear, Chapter 5.3.2. The experiments cover a wide range of parameters such as specimen size, reinforcement ratios, reinforcement types (deformed, plain, unbonded) and concrete strength. Principal tensile strains ε_1 at the ultimate state were recorded for 30 of the 78 specimens.

The experiments summarised in Tab. B.1 have been selected from more than 200 tests, eliminating those specimens in which both reinforcements yielded at the ultimate state. As outlined in Chapter 5.3.2, collapse eventually occurs by crushing of the concrete even in cases where both reinforcements yield at the ultimate state (unless rupture of the reinforcement is governing). Basically, such tests could be included in the comparison of compression softening relationships with the experimental evidence [156,57]; however, they have been excluded here for two reasons. First, the concrete compressive strength is of little practical interest in these cases since the ultimate load for which a panel has been designed was attained if both reinforcements yielded. Second, the value of f_c is determined by the reinforcement ratios rather than by the concrete itself if both reinforcements yield at the ultimate state. Neglecting the concrete tensile strength, the principal concrete compressive stresses are equal to $-\sigma_{c3} = (\rho_x \sigma_{sx} - \sigma_x) + (\rho_z \sigma_{sz} - \sigma_z)$, see Fig. 4.6 (c), and thus, f_c cannot exceed a value of $(\rho_x f_{sux} - \sigma_x) + (\rho_z f_{suz} - \sigma_z)$.

The loading rates used in determining yield strengths f_{sy} are typically higher than those occurring in panel tests; however, the value of f_{sy} increases with the rate of loading. For example, the values of f_{sy} measured at a stress rate of 10 MPa s^{-1} typically exceed those obtained for static loading by about 5 % [65]. Hence, in order to eliminate panels in which both reinforcements yielded at ultimate, only specimens that failed at applied shear stresses below 95 % of the corresponding ultimate load have been included in Tab. B.1, $\tau_u < 0.95 \sqrt{\tilde{\rho}_x f_{syx} \tilde{\rho}_z f_{syz}}$, see Eq. (5.33)$_1$.

In Tab. B.1, the values of $\tau_{u\,calc}$ in the columns labelled '.5 f_c'', '(2.25)' and '(2.27)' have been calculated from Eqs. (4.4)$_2$ and (4.4)$_3$, using the values of f_c obtained from the expression $f_c = 0.5 f_c'$, from Eq. (2.25), and from Eq. (2.27), respectively. The values of $\tau_{u\,calc}$ according to Eqs. (2.25) and (2.27) could only be determined for the first 30 panels, for which principal tensile strains ε_1 recorded at the ultimate state were available. The value of $\tau_{u\,calc}$ in the last column in Tab. B.1 has been calculated directly from Eqs. (5.33)$_2$ and (5.33)$_3$. By rearranging Eqs. (4.4)$_2$ and (4.4)$_3$ the principal concrete compressive stresses at the ultimate state can be calculated from $\tau_{u\,exp}$; the values of f_c for the first 30 panels obtained in this way have been used in plotting Fig. 5.6 (c). A statistical analysis reveals that the values of ε_1 and f_c' influence the ratio f_c / f_c' to a comparable extent; the corresponding coefficients of correlation amount to -0.720 and -0.673, respectively.

From Tab. B.1 it can be seen that both Eq. (2.27) and Eq. (5.33) correlate well with the experimental evidence; the corresponding coefficients of variation are significantly below those obtained from both the assumption $f_c = 0.5 f_c'$ and from Eq. (2.25). This is due to the fact that, as outlined above, the values of ε_1 and f_c' influence the ratio f_c / f_c', whereas the assumption $f_c = 0.5 f_c'$ neglects both influences and Eq. (2.25) accounts only for the influence of ε_1. Note that the coefficients of variation for the assumption $f_c = 0.5 f_c'$ and for Eq. (5.33) amount to 0.302 and 0.145, respectively, if only the first 30 panels are considered.

Source	Specimen	$b=h$	t	f'_c	ρ_x	ρ_z	f_{svx}	f_{svz}	τ_{uexp}	σ_{cu}	σ_{zu}	ε_{1u}	τ_{uexp}/τ_{ucalc} [-]			
[-]	[-]	[mm]	[mm]	[MPa]	[%]	[%]	[MPa]	[MPa]	[MPa]	[MPa]	[MPa]	[‰]	$.5\,f'_c$	(2.25)	(2.27)	(5.33)
[155]	PV9	890	70	11.6	1.785	1.785	455	455	3.76	–	–	5.12	1.30	1.08	0.81	0.85
	PV19	890	70	19.0	1.785	0.713	458	299	3.96	–	–	11.32	1.00	1.23	0.99	1.06
	PV20	890	70	19.6	1.785	0.885	460	297	4.26	–	–	7.21	0.98	0.99	0.87	1.01
	PV21	890	70	19.5	1.785	1.296	458	302	5.03	–	–	8.10	1.05	1.13	0.94	0.97
	PV22	890	70	19.6	1.785	1.524	458	420	6.07	–	–	4.23	1.24	0.94	0.88	0.97
	PV23	890	70	20.5	1.785	1.785	518	518	8.88	-3.50 [a]	-3.50 [a]	4.80	1.73	1.40	1.29	1.37
	PV25	890	70	19.3	1.785	1.785	466	466	9.13	-6.35 [a]	-6.35 [a]	2.87	1.90	1.22	1.24	1.47
	PV27	890	70	20.5	1.785	1.785	442	442	6.35	–	–	3.42	1.24	0.86	0.85	0.98
	PV28	890	70	19.0	1.785	1.785	483	483	5.80	1.80 [a]	1.80 [a]	4.57	1.22	0.96	0.88	0.94
[158]	PHS3	890	70	58.4	3.230	0.820	606	521	8.19	–	–	8.53	0.79	0.85	0.92	1.07
	PHS7	890	70	53.6	3.230	0.820	606	521	10.26	-2.57 [a]	-2.57 [a]	5.79	0.87	0.81	0.92	1.08
	PHS9	890	70	56.0	3.230	0.410	606	521	9.37	-2.34 [a]	-2.34 [a]	12.39	0.95	1.19	1.20	1.31
[9]	TP1A	890	70	25.6	2.040	1.020	450	450	5.75	–	–	4.75	0.94	0.80	0.79	0.93
	KP1	2510	140	25.2	2.040	1.020	430	430	5.62	–	–	4.91	0.94	0.81	0.80	0.94
	KP4	2510	140	24.1	2.040	2.040	430	430	6.88	–	–	3.83	1.14	0.83	0.85	0.96
[116]	SM20	2510	140	28.1	2.040	2.040	398	398	7.23	–	–	5.74	1.03	0.91	0.90	0.91
[68] [70]	SE1	1524	285	42.5	2.930	0.978	492	479	6.76	–	–	8.77	0.77	0.84	0.84	0.92
	SE5	1524	285	25.9	2.930	2.930	492	492	8.11	–	–	10.77	1.25	1.65	1.34	1.07
	SE6	1524	285	40.0	2.930	0.326	492	479	3.75	–	–	10.23	0.70	0.80	0.78	1.01
[31]	SE11	1524	285	70.8	2.930	0.980	480	451	6.56	–	–	12.88	0.56	0.71	0.75	0.80
	SE12	1524	285	75.9	2.930	0.980	480	451	7.41	–	–	14.70	0.61	0.82	0.86	0.89
	SE13	1524	285	80.5	6.400	1.920	510	480	11.98	–	–	10.98	0.71	0.86	0.98	0.94
[91]	PP1	1524	285	27.0	1.942	0.647	479	480	4.95	–	–	11.51	0.87	1.09	0.94	0.97
[117]	A4	1397	178	42.5	2.840	2.840	470	470	11.31	–	–	6.50	1.07	1.02	1.11	1.08
	B4	1397	178	44.8	2.840	0.571	470	445	5.07	–	–	13.17	0.71	0.91	0.87	0.98
	B5	1397	178	42.8	2.840	1.143	470	463	7.16	–	–	10.15	0.77	0.91	0.90	0.91
	B6	1397	178	43.0	2.840	1.710	470	446	9.15	–	–	8.95	0.89	1.00	1.01	0.98
[170]	VA4	1397	203	103.1	4.990	4.990	470	470	21.40	–	–	6.41	0.83	0.78	1.15	1.13
	VB3	1397	178	102.3	5.700	1.143	470	445	9.71	–	–	9.33	0.63	0.70	0.83	0.99
[167]	S-21	1200	200	19.0	4.280	4.280	378	378	6.52	–	–	–	1.37	–	–	1.06
	S-31	1200	200	30.2	4.280	4.280	378	378	8.34	–	–	–	1.10	–	–	1.00
	S-32	1200	200	30.8	3.380	3.380	381	381	8.76	–	–	–	1.14	–	–	1.03
	S-33	1200	200	31.4	2.580	2.580	392	392	8.17	–	–	–	1.04	–	–	0.95
	S-34	1200	200	34.6	1.910	1.910	418	418	7.35	–	–	–	0.85	–	–	0.80
	S-41	1200	200	38.7	4.280	4.280	409	409	11.87	–	–	–	1.23	–	–	1.20
	S-42	1200	200	38.7	4.280	4.280	409	409	12.75	–	–	–	1.32	–	–	1.29
	S-43	1200	200	41.0	4.280	4.280	409	409	11.87	–	–	–	1.16	–	–	1.16
	S-44	1200	200	41.0	4.280	4.280	409	409	12.16	–	–	–	1.19	–	–	1.19
	S-61	1200	200	60.7	4.280	4.280	409	409	15.40	–	–	–	1.01	–	–	1.16
	S-62	1200	200	60.7	4.280	4.280	409	409	15.60	–	–	–	1.03	–	–	1.17
	S-81	1200	200	79.7	4.280	4.280	409	409	16.19	–	–	–	0.81	–	–	1.01
	S-82	1200	200	79.7	4.280	4.280	409	409	16.28	–	–	–	0.82	–	–	1.02
[148]	SP1.7/0	600	50	20.9	1.750	1.750	444	444	6.07	–	–	–	1.16	–	–	0.93
	SP2.4/0	600	50	24.5	2.440	2.440	444	444	6.16	–	–	–	1.00	–	–	0.85
	SP1.0/50	600	50	17.8	1.010	1.010	444	444	5.05	-4.91 [b]	–	–	1.14	–	–	0.95
	SP1.7/50	600	50	19.5	1.750	1.750	444	444	6.05	-4.91 [b]	–	–	1.24	–	–	0.97
	SP2.4/50	600	50	19.8	2.440	2.440	444	444	7.09	-4.91 [b]	–	–	1.43	–	–	1.12
	SP1.0/100	600	50	21.6	1.010	1.010	444	444	6.27	-9.81 [b]	–	–	1.18	–	–	1.09
	SP1.7/100	600	50	22.9	1.750	1.750	444	444	6.77	-9.81 [b]	–	–	1.18	–	–	0.97
	SP2.4/100	600	50	21.1	2.440	2.440	444	444	7.80	-9.81 [b]	–	–	1.48	–	–	1.19
	SP1.7/-25	600	50	18.2	1.750	1.750	444	444	5.54	–	2.45 [b]	–	1.22	–	–	0.97
	SP2.4/-25	600	50	18.0	2.440	2.440	444	444	5.77	–	2.45 [b]	–	1.29	–	–	0.98
[149]	A-3	600	80	21.1	2.000	2.000	401	401	7.15	–	–	–	1.36	–	–	1.09
	A-5	600	80	21.6	1.470	1.470	401	401	5.46	–	–	–	1.01	–	–	0.82
	A-6	600	80	24.0	2.000	2.000	401	401	5.57	–	–	–	0.93	–	–	0.78
	A-9	600	80	23.3	2.000	2.000	401	401	7.05	–	–	–	1.21	–	–	1.00
	B-7	600	80	22.7	1.470	1.470	401	401	5.55	–	–	–	0.98	–	–	0.80
	B-13	600	80	31.1	2.000	2.000	401	401	6.33	–	–	–	0.81	–	–	0.74
	B-14	600	80	31.1	2.000	2.000	401	401	5.50	–	–	–	0.71	–	–	0.65
	B-15	600	80	31.1	2.000	2.000	401	401	6.77	–	–	–	0.87	–	–	0.79
	B-16	600	80	30.6	2.000	2.000	401	401	6.25	–	–	–	0.82	–	–	0.74
	C-1	600	80	31.4	1.470	0.730	401	401	3.59	–	–	–	0.59	–	–	0.70
	C-2	600	80	31.2	1.470	0.730	401	401	3.76	–	–	–	0.62	–	–	0.73
	C-5	600	80	23.0	2.000	1.060	401	401	5.47	–	–	–	0.99	–	–	0.96
[100]	A8-15-8	600	80	96.1	1.863	1.863	922	922	15.91	–	–	–	0.66	–	–	0.88
	A8-20-8	600	80	98.2	2.474	2.474	922	922	17.70	–	–	–	0.72	–	–	0.96
	A8-25-8/1	600	80	98.7	3.105	3.105	922	922	19.33	–	–	–	0.78	–	–	1.05
	A8-25-8/2	600	80	98.7	3.105	3.105	922	922	19.69	–	–	–	0.80	–	–	1.07
	A8-20/8-8	600	80	98.2	2.474	0.990	922	922	11.56	–	–	–	0.60	–	–	0.85
	B5-15-8	600	80	72.6	1.863	1.863	922	922	15.89	–	–	–	0.88	–	–	1.06
	B5-20-8	600	80	72.6	2.474	2.474	922	922	16.66	–	–	–	0.92	–	–	1.11
	B5-25-8	600	80	68.5	3.105	3.105	922	922	17.14	–	–	–	1.00	–	–	1.19
	C3-20-8	600	80	42.5	2.474	2.474	922	922	14.95	–	–	–	1.41	–	–	1.42
[163]	PL45-DII	600	60	30.9	2.610	2.610	318	318	7.76	–	–	–	1.00	–	–	0.91
	PL45-PCI	600	60	30.4	0.775	0.775	1187	1187	7.93	–	–	–	1.04	–	–	0.94
	PL45-PCII	600	60	30.4	1.550	1.550	1187	1187	11.72	–	–	–	1.54	–	–	1.40
	PL45-PCIII	600	60	30.4	1.550	0.775	1187	1187	9.44	–	–	–	1.24	–	–	1.12
	PL45-PCIV	600	60	44.9	1.550	0.775	1187	1187	10.63	–	–	–	0.96	–	–	1.03

[a] axial stresses proportional to applied shear stress.
[b] axial stresses applied first, constant during test.

Average τ_{uexp}/τ_{ucalc}	1.019	0.969	0.947	1.005
Coefficient of variation	0.261	0.220	0.168	0.163

Tab. B.1 – Panels failing by crushing of the concrete.

References

[1] ACI-ASCE Committee 326, "Shear and Diagonal Tension," *ACI Journal*, Vol. 59, No 1, 2, and 3, Jan., Feb., and March 1962, pp. 1-30, 277-344, and 352-396.

[2] ACI-ASCE Committee 426, "The Shear Strength of Reinforced Concrete Members," *Journal of the Structural Division*, ASCE, Vol. 99, No. ST6, June 1973, pp. 1091-1187.

[3] ACI-ASCE Committee 426, "Shear in Reinforced Concrete," *ACI Special Publication* SP-42, Detroit, 1974, Vol. 1, pp. 1-424, and Vol. 2, pp. 425-949.

[4] ACI Committee 318, *Building Code Requirements for Structural Concrete (ACI 318-95) and Commentary ACI 318R-95*, American Concrete Institute, Detroit, 1995, 369 pp.

[5] Adebar, P., and Collins, M.P., "Shear Strength of Members Without Transverse Reinforcement," *Canadian Journal of Civil Engineering*, Vol. 23, No. 1, Jan. 1996, pp. 30-41.

[6] Alvarez, M., "Einfluss des Verbundverhaltens auf das Verformungsvermögen von Stahlbeton (Influence of Bond Behaviour on the Deformation Capacity of Structural Concrete)," Institut für Baustatik und Konstruktion, ETH Zürich, *IBK Bericht*, Birkhäuser Verlag, Basel, 1998, in press.

[7] Alvarez, M., and Sigrist, V., "Stress Field Analysis," Comité Euro-International du Béton, *CEB Bulletin d'information*, 1998, in press.

[8] Anderheggen, E., Despot, Z., Steffen, P., and Tabatabai, S.M., "Computer-Aided Dimensioning of Reinforced Concrete Wall and Flat Slab Structures," *Structural Engineering International*, IABSE, Vol. 4, No. 1, Feb. 1994, pp. 17-22.

[9] André, H.M.O., "Toronto / Kajima Study on Scale Effects in Reinforced Concrete Elements," *Ph.D. Thesis*, University of Toronto, Toronto, 1987, 267 pp.

[10] Baumann, T., "Zur Frage der Netzbewehrung von Flächentragwerken (On the Problem of Mesh Reinforcement of Surface Structures)," *Bauingenieur*, Vol. 47, 1972, pp. 367-377.

[11] Bazant, Z.P., and Gambarova, P., "Rough Cracks in Reinforced Concrete," *Journal of the Structural Division*, ASCE, Vol. 106, No. ST4, April 1980, pp. 819-842.

[12] Belarbi, A., and Hsu, T.T.C., "Constitutive Laws of Softened Concrete in Biaxial Tension-Compression," *ACI Structural Journal*, Vol. 92, No. 5, Sept.-Oct. 1995, pp. 562-573.

[13] Bhide, S.B., and Collins, M.P., "Reinforced Concrete Elements in Shear and Tension," University of Toronto, Department of Civil Engineering, *Publication* No. 87-02, Toronto, Jan. 1987, 332 pp.

[14] Bhide, S.B., and Collins, M.P., "Influence of Axial Tension on the Shear Capacity of Reinforced Concrete Members," *ACI Structural Journal*, Vol. 86, No. 5, Sept.-Oct. 1989, pp. 570-581.

[15] Birkeland, P.W., and Birkeland, H.W., "Connections in Precast Concrete Construction," *ACI Journal*, Vol. 63, No. 3, March 1966, pp. 345-368.

[16] Braestrup, M.W., "Plastic Analysis of Shear in Reinforced Concrete," *Magazine of Concrete Research*, Vol. 26, No. 89, Dec. 1974, pp. 221-228.

[17] Brenni, P., "Il comportamento al taglio di una struttura a sezione mista in calcestruzzo a getti successivi (Shear Behaviour of Segmentally Cast Concrete Structures)," Institut für Baustatik und Konstruktion, ETH Zürich, *IBK Bericht* Nr. 211, Birkhäuser Verlag, Basel, Sept. 1995, 150 pp.

[18] Bulicek, H., "Zur Berechnung des ebenen Spannungs- und des Verzerrungszustandes von schubbewehrten Stegen profilierter Stahlbeton- und Spannbetonträger im Grenzzustand der Schubtragfähigkeit (On the Determination of the States of Stress and Strain in the Webs of Structural Concrete Girders with Shear Reinforcement in the Ultimate Limit State)," *Berichte aus dem Konstruktiven Ingenieurbau*, Technische Universität München, Heft 4, 1993, 126 pp.

[19] Bulicek, H., und Kupfer, H., "Ebener Spannungs- und Verzerrungszustand von schubbewehrten Stegen profilierter Stahlbeton- und Spannbetonträger im Grenzzustand der Schubtragfähigkeit (Plane States of Stress and Strain in the Webs of Structural Concrete Girders with Shear Reinforcement in the Ultimate Limit State)," *Beiträge zum 28. Forschungskolloquium*, Deutscher Ausschuss für Stahlbeton, München, 1993, pp. 76-84.

[20] Canadian Standards Association, *Design of Concrete Structures for Buildings (CAN3-A23.3-M84)*, Rexdale, Ontario, Canada, 1984, 281 pp.

[21] Cerruti, L., and Marti, P., "Staggered Shear Design of Concrete Beams: Large Scale Tests," *Canadian Journal of Civil Engineering*, Vol. 14, No. 2, April 1987, pp. 257-268.

[22] Chen, W.F., and Drucker, D.C., "Bearing Capacity of Concrete Blocks or Rock," *Proceedings*, ASCE, Engineering Mechanics Division, Vol. 95, No. 4, 1969, pp. 955-978.

[23] Chen, W.F., "Double Punch Test for Tensile Strength of Concrete," *ACI Journal*, Vol. 67, No. 12, Dec. 1970, pp. 993-995.

[24] Chen, W.F., *Limit Analysis and Soil Plasticity*, Developments in Geotechnical Engineering, Vol. 7, Elsevier, Amsterdam, 1975, 638 pp.

[25] Chen, W.F., *Plasticity in Reinforced Concrete*, Mac Graw-Hill, New York, 1982, 484 pp.

[26] Chen, R.C., Carrasquillo, R.L., and Fowler, D.W., "Behavior of High-Strength Concrete Under Uniaxial and Biaxial Compression," *ACI Special Publication* SP-87, Detroit, 1985, pp. 251-274.

[27] Collins, M.P., "Towards a Rational Theory for RC Members in Shear," *Journal of the Structural Division*, ASCE, Vol. 104, No. ST4, April 1978, pp. 649-666.

[28] Collins, M.P., "Stress-Strain Characteristics of Diagonally Cracked Concrete," IABSE Colloquium 'Plasticity in Reinforced Concrete', Copenhagen 1979, International Association for Bridge and Structural Engineering, *Final Report*, IABSE Vol. 29, 1979, pp. 27-34.

[29] Collins, M.P., and Mitchell, D., "Shear and Torsion Design of Prestressed and Non-Prestressed Concrete Beams," *PCI Journal*, Vol. 25, No. 5, Sept.-Oct. 1980, pp. 32-100.

[30] Collins, M.P., Mitchell, D., and Mehlhorn, G., "An International Competition to Predict the Response of Reinforced Concrete Panels," *Canadian Journal of Civil Engineering*, Vol. 12, No. 3, Sept. 1985, pp. 624-644.

[31] Collins, M.P., and Porasz, A., "Shear Design for High-Strength Concrete," Comité Euro-International du Béton, *CEB Bulletin d'information,* No. 193, Dec. 1989, pp. 77-83.

[32] Collins, M.P., and Mitchell, D., *Prestressed Concrete Structures*, Prentice-Hall, Englewood Cliffs, NJ, 1991, 766 pp.

[33] Collins, M.P., Mitchell, D., Adebar, P., and Vecchio, F.J., "A General Shear Design Method," *ACI Structural Journal*, Vol. 93, No. 1, Jan.-Feb. 1996, pp. 36-45.

[34] Comité Euro-International du Béton, *CEB-FIP Model Code for Concrete Structures*, Third Edition, Paris, 1978, 348 pp.

[35] Comité Euro-International du Béton, *CEB-FIP Model Code 1990*, First Edition, London, 1993, 437 pp.

[36] Cornelissen, H.A.W., Hordijk, D.A., and Reinhardt, H.W., "Experimental Determination of Crack Softening Characteristics of Normal Weight and Lightweight Concrete," *Heron*, Vol. 31, No. 2, Delft University of Technology, 1986, pp. 45-56.

[37] Courant, R., und Hilbert, D., *Methoden der mathematischen Physik II* (Methods of Mathematical Physics II), Springer Verlag, Berlin, 1937, 549 pp.

[38] Daschner, F., und Kupfer, H., "Versuche zur Schubkraftübertragung in Rissen von Normal- und Leichtbeton (Tests on Shear Transfer Across Cracks in Normal and Lightweight Concrete)," *Bauingenieur*, Vol. 57, 1982, pp. 57-60.

[39] Dei Poli, S., Gambarova, P.G., and Karakoç, C., "Aggregate Interlock Role in RC Thin-Webbed Beams in Shear," *Journal of Structural Engineering*, ASCE, Vol. 113, No. 1, Jan. 1987, pp. 1-19.

[40] Dei Poli, S., di Prisco, M., and Gambarova, P.G., "Stress Field in Web of RC Thin-Webbed Beams Failing in Shear," *Journal of Structural Engineering*, ASCE, Vol. 116, No. 9, Sept. 1990, pp. 2497-2515.

[41] Delhumeau, G., Gubler, J., Legault, R., and Simonnet, C., *Le béton en représentation, la mémoire photographique de l'entreprise Hennebique* (Representation of Concrete, the Photographic Memory of the Hennebique Company), Editions Hazan, Paris, France, 1993, 190 pp.

[42] Drucker, D.C., Greenberg, H.J., and Prager, W., "The Safety Factor of an Elastic-Plastic Body in Plane Strain," *Journal of Applied Mechanics*, ASME, Vol. 18, 1951, pp. 371-378.

[43] Drucker, D.C., Greenberg, H.J., Lee, E.H., and Prager, W., "On Plastic-Rigid Solutions and Limit Design Theorems for Elastic-Plastic Bodies," *Proceedings,* First US National Congress of Applied Mechanics, Chicago, Illinois, June 11-16 1951, ASME, 1952, pp. 533-538.

[44] Drucker, D.C., Greenberg, H.J., and Prager, W., "Extended Limit Design Theorems for Continuous Media," *Quarterly of Applied Mathematics*, Vol. 9, 1952, pp. 381-389.

[45] Drucker, D.C., "The Effect of Shear on the Plastic Bending of Beams," *Journal of Applied Mechanics*, ASME, Vol. 23, 1956, pp. 509-514.

[46] Eibl, J., und Neuroth, U., "Untersuchungen zur Druckfestigkeit von bewehrtem Beton bei gleichzeitig wirkendem Querzug (On the Compressive Strength of Reinforced Concrete Subjected to Lateral Tension)," *Forschungsbericht* T2024, Institut für Massivbau und Baustofftechnologie, Universität Karlsruhe, 1988, 136 pp.

[47] Engesser, F., "Über statisch unbestimmte Träger bei beliebigem Formänderungs-Gesetze und über den Satz von der kleinsten Ergänzungsarbeit (On Statically Indeterminate Girders with Arbitrary Stress-Strain Relationship and the Principle of Minimum Complementary Energy)," *Zeitschrift des Architekten- und Ingenieur-Vereins zu Hannover*, Band 25, Heft 8, 1889, Spalten 733-744.

[48] *Eurocode 2: Design of Concrete Structures*, 'Part 1-1: General Rules and Rules for Buildings (ENV 1992-1-1)', Comité Européen de Normalisation, Brussels, Belgium, 1991, 173 pp.

[49] Fédération Internationale de la Précontrainte, "High Strength Concrete," *State of the Art Report*, No. SR 90/1, London, Aug. 1990, 61 pp.

[50] Fenwick, R.C., and Paulay, T., "Mechanisms of Shear Resistance of Concrete Beams," *Journal of the Structural Division*, ASCE, Vol. 94, No. ST10, Oct. 1968, pp. 2325-2350.

[51] Gvozdev, A.A., "The Determination of the Value of the Collapse Load for Statically Indeterminate Systems Undergoing Plastic Deformation," *International Journal of Mechanical Sciences*, Vol. 1, 1960, pp. 322-335 (English translation of the Russian original published in *Proceedings* of the Conference on Plastic Deformations, December 1936, Akademiia Nauk SSSR, Moscow-Leningrad, 1938, pp. 19-38).

[52] Hill, R., "On the State of Stress in a Plastic-Rigid Body at the Yield Point," *The Philosophical Magazine*, Vol. 42, 1951, pp. 868-875.

[53] Hillerborg, A, "Analysis of a Single Crack," *Fracture Mechanics of Concrete*, edited by F.H. Wittmann, Elsevier, Amsterdam, 1983, pp. 223-249.

[54] Hodge, P.G. Jr., "Interaction Curves for Shear and Bending of Plastic Beams," *Journal of Applied Mechanics*, ASME, Vol. 24, 1957, pp. 453-456.

[55] Hofbeck, J.A., Ibrahim, I.O., and Mattock, A.H., "Shear Transfer in Reinforced Concrete," *ACI Journal*, Vol. 66, No. 2, Feb. 1969, pp. 119-128.

[56] Hori, E., "Size Effect on Concentric Compression Characteristics of High Strength Reinforced Concrete Columns Confined by Lateral Reinforcement (in Japanese)," *Summaries of Technical Papers of Annual Meeting*, Architectural Institute of Japan, Structures II, Vol. C, 1994, pp. 365-366.

[57] Hsu, T.T.C., "Softened Truss Model Theory for Shear and Torsion," *ACI Structural Journal*, Vol. 85, No. 6, Nov.-Dec. 1988, pp. 624-635.

[58] Hsu, T.T.C., Belarbi, A., and Pang, X., "A Universal Panel Tester," Department of Civil and Environmental Engineering, University of Houston, *Research Report* UHCEE 91-1, Houston, July 1991, 48 pp.

[59] Hsu, T.T.C., and Zhang, L.-X., "Tension Stiffening in Reinforced Concrete Membrane Elements," *ACI Structural Journal*, Vol. 93, No. 1, Jan.-Feb. 1996, pp. 108-115.

[60] IABSE Colloquium 'Plasticity in Reinforced Concrete', Copenhagen 1979, International Association for Bridge and Structural Engineering, *Introductory Report*, IABSE Vol. 28, 1978, 172 pp., and *Final Report*, IABSE Vol. 29, 1979, 360 pp.

[61] Kanellopoulos, A., "Zum unelastischen Verhalten und Bruch von Stahlbeton (On the Inelastic Behaviour and Failure of Structural Concrete)," Institut für Baustatik und Konstruktion, ETH Zürich, IBK Bericht Nr. 153, Birkhäuser Verlag, Basel, Nov. 1986, 86 pp.

[62] Kani, G.N.J., "The Riddle of Shear Failure and Its Solution," *ACI Journal*, Vol. 61, No. 4, April 1964, pp. 441-462.

[63] Kani, M.W., Huggins, M.W., and Wittkopp, R.R., "Kani on Shear in Reinforced Concrete," *Memorial Volume* in honour of G.N.J. Kani, Department of Civil Engineering, University of Toronto, 1979, 225 pp.

[64] Karihaloo, B.L., *Fracture Mechanics and Structural Concrete*, Longman Scientific & Technical, Concrete Design and Construction Series, Essex, 1995, 330 pp.

[65] Kaufmann, W., und Marti, P., "Versuche an Stahlbetonträgern unter Normal- und Querkraft (Tests on Structural Concrete Girders Subjected to Shear and Normal Forces)," Institut für Baustatik und Konstruktion, ETH Zürich, *IBK Bericht* Nr. 226, Birkhäuser Verlag, Basel, Nov. 1996, 131 pp.

[66] Kaufmann, W., "Large-Scale Tests on Structural Concrete Girders under Shear and Normal Forces," *Proceedings*, Ph.D. Symposium, Technical University of Budapest, May 1996, pp. 17-21.

[67] Kaufmann, W., and Marti, P., "Structural Concrete: Cracked Membrane Model," *Journal of Structural Engineering*, ASCE, 1998, in press.

[68] Khalifa, J., "Limit Analysis and Design of Reinforced Concrete Shell Elements," *Ph.D. Thesis*, University of Toronto, Toronto, 1986, 314 pp.

[69] Kirmair, H., und Mang, R., "Das Tragverhalten der Schubzone schlanker Stahlbeton- und Spannbetonträger bei Biegung mit Längskraft (Behaviour of the Web of Slender Structural Concrete Girders Subjected to Flexure and Axial Forces)," *Bauingenieur*, Vol. 62, 1987, pp. 165-170.

[70] Kirschner, U., and Collins, M.P., "Investigating the Behaviour of Reinforced Concrete Shell Elements," University of Toronto, Department of Civil Engineering, *Publication* No. 86-09, Toronto, Sept. 1986, 210 pp.

[71] Kollegger, J. und Mehlhorn, G., "Experimentelle Untersuchungen zur Bestimmung der Druckfestigkeit des gerissenen Stahlbetons bei einer Querzugbeanspruchung (Experimental Investigations for the Determination of the Compressive Strength of Cracked Reinforced Concrete Subjected to Lateral Tension)," Deutscher Ausschuss für Stahlbeton, *Heft* 413, 1990, 132 pp.

[72] Kupfer, H., "Erweiterung der Mörsch'schen Fachwerkanalogie mit Hilfe des Prinzips vom Minimum der Formänderungsenergie (Generalisation of Mörsch's Truss Analogy Using the Principle of Minimum Strain Energy)," Comité Euro-International du Béton, *CEB Bulletin d'information*, No. 40, Jan. 1964, pp. 44-57.

[73] Kupfer, H., "Das Verhalten des Betons unter zweiachsiger Beanspruchung (Biaxial Behaviour of Concrete)," *Bericht*, Lehrstuhl für Massivbau, Technische Universität München, No. 78, 1969, 124 pp.

[74] Kupfer, H., Mang, R., und Karavesyroglou, M., "Bruchzustand der Schubzone von Stahlbeton- und Spannbetonträgern – Eine Analyse unter Berücksichtigung der Rissverzahnung (Ultimate Limit State of the Web of Structural Concrete Girders – An Analysis Accounting for Aggregate Interlock)," *Bauingenieur*, Vol. 58, 1983, pp. 143-149.

[75] Kupfer, H., and Bulicek, H., "A Consistent Model for the Design of Shear Reinforcement in Slender Beams with I- or Box-Shaped Cross-Section," *Proceedings*, International Workshop 'Concrete Shear in Earthquake', Houston, Texas, Jan. 13-16, 1991, Elsevier, 1992, pp. 256-265.

[76] Lampert, P., and Collins, M.P., "Torsion, Bending and Confusion – An Attempt to Establish the Facts," *ACI Journal*, Vol. 69, No. 8, Aug. 1972, pp. 500-504.

[77] Lüthi, M., "Neuauswertung von Schubversuchen an profilierten Trägern (Re-evaluation of Shear Tests on Profiled Girders)," *Diplomarbeit*, Institut für Baustatik und Konstruktion, ETH Zürich, Jan. 1997, 95 pp.

[78] Maier, J., "Tragfähigkeit von Stahlbetonscheiben (Load-Bearing Capacity of Structural Concrete Walls)," Institut für Baustatik und Konstruktion, ETH Zürich, *IBK Bericht* Nr. 169, Birkhäuser Verlag, Basel, Nov. 1988, 93 pp.

[79] Markeset, G., "Size Effect on Stress-Strain Relationship of Concrete in Compression," *Proceedings*, Symposium 'Utilization of High Strength Concrete', Lillehammer, 1993, Vol. 2, pp. 1146-1153.

[80] Marti, P., "Plastische Berechnungsmethoden (Limit Analysis Methods)," *Vorlesungsunterlagen*, Abteilung für Bauingenieurwesen, ETH Zürich, April 1978, 82 pp.

[81] Marti, P., "Plastic Analysis of Reinforced Concrete Shear Walls," IABSE Colloquium 'Plasticity in Reinforced Concrete', Copenhagen 1979, International Association for Bridge and Structural Engineering, *Introductory Report*, IABSE Vol. 28, 1978, pp. 51-69.

[82] Marti, P., "Zur Plastischen Berechnung von Stahlbeton (Limit Analysis of Structural Concrete)," Institut für Baustatik und Konstruktion, ETH Zürich, *IBK Bericht* Nr. 104, Birkhäuser Verlag, Basel, Oct. 1980, 176 pp.

[83] Marti, P., "Strength and Deformations of Reinforced Concrete Members under Torsion and Combined Actions," Comité Euro-International du Béton, *CEB Bulletin d'information,* No. 146, 1982, pp. 97-138.

[84] Marti, P., "Basic Tools of Reinforced Concrete Beam Design," *ACI Journal*, Vol. 82, No. 1, Jan.-Feb. 1985, pp. 46-56.

[85] Marti, P., "Truss Models in Detailing," *Concrete International*, Vol. 7, No. 12, Dec. 1985, pp. 66-73.

[86] Marti, P., "Staggered Shear Design of Simply Supported Concrete Beams," *ACI Structural Journal*, Vol. 83, No. 1, Jan.-Feb. 1986, pp. 36-42.

[87] Marti, P., "Staggered Shear Design of Concrete Bridge Girders," *Proceedings*, International Conference on Short and Medium Span Bridges, Vol. 1, Ottawa, Aug. 1986, pp. 139-149.

[88] Marti, P., "Design of Concrete Slabs for Transverse Shear," *ACI Structural Journal*, Vol. 87, No. 2, March-April 1990, pp. 180-190.

[89] Marti, P., "Dimensioning and Detailing," IABSE Colloquium 'Structural Concrete', Stuttgart 1991, International Association for Bridge and Structural Engineering, *Report*, IABSE Vol. 62, 1991, pp. 411-443.

[90] Marti, P., "State-of-the-Art of Membrane Shear Behavior – European Work," *Proceedings*, International Workshop 'Concrete Shear in Earthquake', Houston, Texas, Jan. 13-16, 1991, Elsevier, 1992, pp. 187-195.

[91] Marti, P., and Meyboom, J., "Response of Prestressed Concrete Elements to In-Plane Shear Forces," *ACI Structural Journal*, Vol. 89, No. 5, Sept.-Oct. 1992, pp. 503-514.

[92] Marti, P., "Shear Design of Variable Depth Girders with Draped Prestressing Tendons," *FIP Report* 'Prestressed Concrete in Switzerland', 12th FIP Congress, Washington, D.C. USA, May 1994, pp. 16-19.

[93] Marti, P., Sigrist, V., und Alvarez, M., "Mindestbewehrung von Betonbauten (Minimum Reinforcement of Concrete Structures)," *Report No. 529*, Research Grant No. 82/95, Swiss Federal Highway Administration, June 1997, 55 pp.

[94] Marti, P., Alvarez, M., Kaufmann, W., and Sigrist, V., "Tension Chord Model for Structural Concrete," *Structural Engineering International*, IABSE, 1998, in press.

[95] Marti, P., "How to Treat Shear in Structural Concrete," *ACI Structural Journal*, 1998, in press.

[96] Martinez, S., Nilson, A.H., and Slate, F.O., "Spirally Reinforced High-Strength Concrete Columns," *ACI Journal*, Vol. 81, No. 5, Sept.-Oct. 1984, pp. 431-442.

[97] Mast, R.F., "Auxiliary Reinforcement in Concrete Connections," *Journal of the Structural Division*, ASCE, Vol. 94, No. ST6, June 1968, pp. 1485-1504.

[98] Mattock, A.H., "Shear Transfer in Reinforced Concrete – Recent Research," *PCI Journal*, Vol. 17, No. 2, March-April 1972, pp. 55-75.

[99] Mattock, A.H., "Shear Transfer in Concrete having Reinforcement at an Angle to the Shear Plane," *ACI Special Publication* SP-42, Vol. 1, 1974, pp. 17-42.

[100] Matuura, T., and Sumi, K., "In-Plane Pure Shear Loading Tests of High Strength Reinforced Concrete Panels (in Japanese)," *Summaries of Technical Papers of Annual Meeting*, Architectural Institute of Japan, Structures II, Vol. C, 1991, pp. 425-428.

[101] Melan, E., "Der Spannungszustand eines Mises-Henckyschen Kontinuums bei veränderlicher Belastung (The State of Stress in a Mises-Hencky Continuum Subjected to Variable Loading)," *Sitzungsberichte*, Akademie der Wissenschaften in Wien, Mathematisch-naturwissenschaftliche Klasse, Abt. IIa, No. 147, 1938, pp. 73-87.

[102] von Mises, R., "Mechanik der plastischen Formänderung von Kristallen (Mechanics of the Plastic Deformation of Crystals)," *Zeitschrift für angewandte Mathematik und Mechanik*, Vol. 8, 1928, pp. 161-185.

[103] Mitchell, D., and Collins, M.P., "Diagonal Compression Field Theory – A Rational Model for Structural Concrete in Pure Torsion," *ACI Journal*, Vol. 71, No. 8, Aug. 1974, pp. 396-408.

[104] Mörsch, E., *Der Eisenbetonbau – seine Theorie und Anwendung* (Reinforced Concrete Construction – Theory and Application), 3. Auflage, Verlag Konrad Wittwer, Stuttgart, 1908, 376 pp.

[105] Mörsch, E., *Der Eisenbetonbau – seine Theorie und Anwendung* (Reinforced Concrete Construction – Theory and Application), 5. Auflage, 1. Band, 1. Hälfte, Verlag Konrad Wittwer, Stuttgart, 1920, 471 pp.

[106] Mörsch, E., *Der Eisenbetonbau – seine Theorie und Anwendung* (Reinforced Concrete Construction – Theory and Application), 5. Auflage, 1. Band, 2. Hälfte, Verlag Konrad Wittwer, Stuttgart, 1922, 460 pp.

[107] Muttoni, A., Schwartz, J., und Thürlimann, B., "Bemessen und Konstruieren von Stahlbetontragwerken mit Spannungsfeldern (Design of Concrete Structures with Stress Fields)," *Vorlesungsunterlagen*, Institut für Baustatik und Konstruktion, ETH Zürich, 1987, 134 pp.

[108] Muttoni, A., "Die Anwendbarkeit der Plastizitätstheorie in der Bemessung von Stahlbeton (Applicability of the Theory of Plasticity to the Design of Structural Concrete)," Institut für Baustatik und Konstruktion, ETH Zürich, *IBK Bericht* Nr. 176, Birkhäuser Verlag, Basel, June 1990, 158 pp.

[109] Müller, P., "Plastische Berechnung von Stahlbetonscheiben und -balken (Plastic Analysis of Reinforced Concrete Disks and Beams)," Institut für Baustatik und Konstruktion, ETH Zürich, *IBK Bericht* Nr. 83, Birkhäuser Verlag, Basel, July 1978, 160 pp.

[110] Neal, P.G., "The Effect of Shear and Normal Forces on the Fully Plastic Moment of a Beam of Rectangular Cross Section," *Journal of Applied Mechanics*, ASME, Vol. 28, 1961, pp. 269-274.

[111] Nelissen, L.J.M., "Biaxial Testing of Normal Concrete," *Heron*, Vol. 18, No. 1, Delft University of Technology, 1972, 90 pp.

[112] Nielsen, M.P., "On the Strength of Reinforced Concrete Discs," Acta Polytechnica Scandinavica, *Civil Engineering and Building Construction Series*, No. 70, Copenhagen, 1971, 261 pp.

[113] Nielsen, M.P., *Limit Analysis and Concrete Plasticity*, Prentice-Hall, Englewood Cliffs, 1984, 420 pp.

[114] Nimura, A., "Experimental Research on Failure Criteria of Ultra-High Strength Concrete under Biaxial Stress (in Japanese)," *Summaries of Technical Papers of Annual Meeting*, Architectural Institute of Japan, Structures II, Vol. C, 1991, pp. 473-474.

[115] Nissen, I., "Rissverzahnung des Betons – gegenseitige Rissuferverschiebungen und übertragene Kräfte (Aggregate Interlock of Concrete – Crack Displacements and Force Transfer)," *Dissertation*, Technische Universität München, 1987, 218 pp.

[116] Ohmori, N., Takahashi, T., Tsubota, H., Inoue, N., Kurihara, K., and Watanabe, S., "Experimental Studies on Non-linear Behaviours of Reinforced Concrete Panels Subjected to Cyclic In-Plane Shear (in Japanese)," *Journal of Structural and Construction Engineering*, Transactions of the Architectural Institute of Japan, No. 403, Sept. 1989, pp. 105-118.

[117] Pang, X., and Hsu, T.T.C., "Constitutive Laws of Reinforced Concrete in Shear," Department of Civil and Environmental Engineering, University of Houston, *Research Report* UHCEE 92-1, Houston, Dec. 1992, 180 pp.

[118] Pang, X., and Hsu, T.T.C., "Fixed Angle Softened Truss Model for Reinforced Concrete," *ACI Structural Journal*, Vol. 93, No. 2, March-April 1996, pp. 197-207.

[119] Pantazopoulou, S.J., "Role of Expansion on Mechanical Behaviour of Concrete," *Journal of Structural Engineering*, ASCE, Vol. 121, No. 12, Dec. 1995, pp. 1795-1805.

[120] Paulay, T., and Loeber, P.J., "Shear Transfer by Aggregate Interlock," *ACI Special Publication* SP-42, Vol. 1, 1974, pp. 1-15.

[121] Potucek, W., "Die Beanspruchung der Stege von Stahlbetonplattenbalken durch Querkraft und Biegung (The State of Stress in the Web of Structural Concrete T-Beams Subjected to Shear and Flexure)," *Zement und Beton*, Vol. 22, No. 3, 1977, pp. 88-98.

[122] Prager, W., *Probleme der Plastizitätstheorie* (Problems of the Theory of Plasticity), Birkhäuser Verlag, Basel, 1955, 100 pp.

[123] Prager, W., "Limit Analysis: the Development of a Concept," *Problems of Plasticity*, edited by A. Sawczuk, Nordhoff International, Leyden, The Netherlands, 1974, pp. 3-24.

[124] Pré, M., "Etude de la torsion dans le béton précontraint par la méthode du treillis spatial évolutif (Study on Torsion in Prestressed Concrete Using the Space Truss Method)," *Annales* de l'Institut Technique du Bâtiment et des Travaux Publics, No. 385, Paris, 1980, pp. 94-112.

[125] di Prisco, M., and Gambarova, P.G., "Comprehensive Model for Study of Shear in Thin-Webbed RC and PC Beams," *Journal of Structural Engineering*, ASCE, Vol. 121, No. 12, Dec. 1995, pp. 1822-1831.

[126] Reineck, K.-H., und Hardjasaputra, H., "Zum Dehnungszustand bei der Querkraftbemessung profilierter Stahlbeton- und Spannbetonträger (On the State of Strain in Shear Design of Profiled Structural Concrete Girders)," *Bauingenieur*, Vol. 65, 1990, pp. 73-82.

[127] Richart, F.E., Brandtzaeg, A., and Brown, R.L., "A Study of the Failure of Concrete under Combined Compressive Stresses," University of Illinois, Engineering Experiment Station, *Bulletin* No. 185, Urbana, Nov. 1928, 103 pp.

[128] Richart, F.E., Brandtzaeg, A., and Brown, R.L., "The Failure of Plain and Spirally Reinforced Concrete in Compression," University of Illinois, Engineering Experiment Station, *Bulletin* No. 190, Urbana, April 1929, 73 pp.

[129] Ritter, W., "Die Bauweise Hennebique (The Hennebique Construction Method)," *Schweizerische Bauzeitung*, Vol. 17, Feb. 1899, pp. 41-43, 49-52 und 59-61.

[130] Robinson, J.R., et Demorieux, J.-M., "Essais de traction-compression sur modèles d'âme de poutre en béton armé (Tension-Compression Tests on Models of the Web of Reinforced Concrete Girders)," Compte rendu des recherches effectuées en 1968, *Annales ITBTP*, June 1969, pp. 980-982, and Compte rendu des recherches effectuées en 1969, *Annales ITBTP*, June 1970, pp. 980-982.

[131] Roos, W., "Zur Druckfestigkeit des gerissenen Stahlbetons in scheibenförmigen Bauteilen bei gleichzeitig wirkender Querzugbelastung (On the Compressive Strength of Cracked Reinforced Concrete in Disks Subjected to Simultaneous Lateral Tension)," *Berichte aus dem Konstruktiven Ingenieurbau*, Technische Universität München, Heft 2, 1995, 174 pp.

[132] Ruina, A.L., "Constitutive Relations for Frictional Slip," *Mechanics of Geomaterials*, John Wiley & Sons Ltd, 1985, pp. 169-188.

[133] Sakino, K., "Experimental Study on RC Columns Confined by High Strength Hoops Under Axial Compression (in Japanese)," *Summaries of Technical Papers of Annual Meeting*, Architectural Institute of Japan, Structures II, Vol. C, 1991, pp. 527-528.

[134] Sakino "Experimental Study on Behavior of RC Columns Confined by High Strength Hoops Under Axially Compressive Load (in Japanese)," *Summaries of Technical Papers of Annual Meeting*, Architectural Institute of Japan, Structures II, Vol. C, 1992, pp. 543-544.

[135] Salençon, J., *Théorie de la plasticité pour les applications à la mécanique des sols* (Theory of Plasticity for Applications in Soil Mechanics), éditions Eyrolles, Paris, 1974, 178 pp.

[136] Sayir, M., und Ziegler, H., "Der Verträglichkeitssatz der Plastizitätstheorie und seine Anwendung auf räumlich unstetige Felder (The Uniqueness Theorem of Limit Analysis and its Application to Spatially Discontinuous Fields)," *Zeitschrift für angewandte Mathematik und Mechanik*, Vol. 20, 1969, pp. 78-93.

[137] Schlaich, J., Schäfer, K., und Schelling, G., "Druck und Querzug in bewehrten Betonelementen (Compression and Lateral Tension in Reinforced Concrete Elements)," *Bericht*, Institut für Massivbau, Universität Stuttgart, Nov. 1982. Also: Deutscher Ausschuss für Stahlbeton, *Heft* 408, 1990, pp. 5-85.

[138] Schlaich, J., Schäfer, K., and Jennewein, M., "Toward a Consistent Design of Structural Concrete," *PCI Journal*, Vol. 32, No. 3, May-June 1987, pp. 74-150.

[139] Selby, R.G., Vecchio, F.J., and Collins, M.P., "Analysis of Reinforced Concrete Members Subject to Shear and Axial Compression," *ACI Structural Journal*, Vol. 93, No. 3, May-June 1996, pp. 306-315.

[140] Setunge, M., Attard, M.M., and Darvall, P.L., "Ultimate Strength of Confined Very High-Strength Concretes," *ACI Structural Journal*, Vol. 90, No. 6, Nov.-Dec. 1993, pp. 632-641.

[141] Shima, H., Chou, L.-L., and Okamura, H., "Micro- and Macro Models for Bond in Reinforced Concrete," *Journal of the Faculty of Engineering*, University of Tokyo, Vol. XXXIX, No. 2, 1987, pp. 133-194.

[142] Sigrist, V., und Marti, P., "Versuche zum Verformungsvermögen von Stahlbetonträgern (Tests on the Deformation Capacity of Structural Concrete Girders)," Institut für Baustatik und Konstruktion, ETH Zürich, *IBK Bericht* Nr. 202, Birkhäuser Verlag, Basel, Nov. 1993, 90 pp.

[143] Sigrist, V., Alvarez, M., and Kaufmann, W., "Shear and Flexure in Structural Concrete Beams," Comité Euro-International du Béton, *CEB Bulletin d'information*, No. 223, June 1995, pp. 7-49.

[144] Sigrist, V., "Zum Verformungsvermögen von Stahlbetonträgern (On the Deformation Capacity of Structural Concrete Girders)," Institut für Baustatik und Konstruktion, ETH Zürich, *IBK Bericht* Nr. 210, Birkhäuser Verlag, Basel, July 1995, 159 pp.

[145] Somes, N.F., "Compression Tests on Hoop-Reinforced Concrete," *Journal of the Structural Division*, ASCE, Vol. 96, No. ST7, July 1970, pp. 1495-1509.

[146] Stoffel, P., and Marti, P., "Structural Evaluation of a 30-Year-Old Post-Tensioned Concrete Bridge," *Proceedings*, FIP Symposium 'Post-Tensioned Concrete Structures', London, Sept. 1996, pp. 267-274.

[147] Stoffel, P., "Zur Beurteilung der Tragsicherheit bestehender Stahlbetonbauten (On the Evaluation of the Strength of Existing Concrete Structures)," Institut für Baustatik und Konstruktion, ETH Zürich, IBK Bericht, Birkhäuser Verlag, Basel, in preparation.

[148] Sumi, K., and Iida, T., "Shear Behavior of Reinforced Concrete Cylindrical Walls (in Japanese)," *Summaries of Technical Papers of Annual Meeting*, Architectural Institute of Japan, Structures II, Vol. C, 1983, pp. 1529-1532.

[149] Sumi, K., "Mechanical Characteristics of Concrete in Reinforced Concrete Panel (in Japanese)," *Summaries of Technical Papers of Annual Meeting*, Architectural Institute of Japan, Structures II, Vol. C, 1987, pp. 365-368.

[150] Teutsch, M., und Kordina, K., "Versuche an Spannbetonbalken unter Kombinierter Beanspruchung aus Biegung, Querkraft und Torsion (Tests on Prestressed Concrete Girders Subjected to Combined Bending, Shear and Torsion)," Deutscher Ausschuss für Stahlbeton, *Heft* 334, 1982, 81 pp.

[151] Thürlimann, B., und Lüchinger, P., "Steifigkeit von gerissenen Stahlbetonbalken unter Torsion und Biegung (Stiffness of Cracked Reinforced Concrete Girders Subjected to Torsion and Bending)," *Beton- und Stahlbetonbau*, Vol. 68, No. 6, June 1973, pp. 146-153.

[152] Thürlimann, B., Marti, P., Pralong, J., Ritz, P., und Zimmerli, B., "Anwendung der Plastizitätstheorie auf Stahlbeton (Application of the Theory of Plasticity to Structural Concrete)," *Unterlagen zum Fortbildungskurs*, Institut für Baustatik und Konstruktion, ETH Zürich, 1983, 252 pp.

[153] Van Mier, J.G.M., "Fracture of Concrete under Complex Stress," *Heron,* Vol. 31, No. 3, Delft University of Technology, 1986, 90 pp.

[154] Vecchio, F.J., and Collins, M.P., "Stress-Strain Characteristics of Reinforced Concrete in Pure Shear," IABSE Colloquium 'Advanced Mechanics of Reinforced Concrete', Delft 1981, International Association for Bridge and Structural Engineering, *Introductory Report*, IABSE Vol. 34, 1981, pp. 211-225.

[155] Vecchio, F.J., and Collins, M.P., "The Response of Reinforced Concrete to In-plane Shear and Normal Stresses," University of Toronto, Department of Civil Engineering, *Publication* No. 82-03, Toronto, March 1982, 332 pp.

[156] Vecchio, F.J., and Collins, M.P., "The Modified Compression Field Theory for Reinforced Concrete Elements Subjected to Shear," *ACI Journal*, Vol. 83, No. 2, March-April 1986, pp. 219-231.

[157] Vecchio, F.J., "Reinforced Concrete Membrane Element Formulations," *Journal of Structural Engineering*, ASCE, Vol. 116, No. 3, March 1990, pp. 730-750.

[158] Vecchio, F.J., Collins, M.P., and Aspiotis, J., "High-Strength Concrete Elements Subjected to Shear," *ACI Structural Journal*, Vol. 91, No. 4, July-August 1994, pp. 423-433.

[159] Vecchio, F.J., and DeRoo, A.,"Smeared-Crack Modelling of Concrete Tension Splitting," *Journal of Engineering Mechanics*, ASCE, Vol. 121, No. 6, June 1995, pp. 702-708.

[160] Wagner, H., "Ebene Blechwandträger mit sehr dünnem Stegblech (Metal Girders with Very Thin Web)," *Zeitschrift für Flugtechnik und Motorluftschiffahrt*, Vol. 20, No. 8, pp. 200-207, No. 9, pp. 227-233, No. 10, pp. 256-262, No. 11, pp. 279-284, and No. 12, pp. 306-314, Berlin, 1929.

[161] Walraven, J.C., "Aggregate Interlock: A Theoretical and Experimental Analysis," *Dissertation*, Delft University Press, 1980, 197 pp.

[162] Walraven, J.C., "Fundamental Analysis of Aggregate Interlock," *Journal of the Structural Division*, ASCE, Vol. 107, No. ST11, Nov. 1981, pp. 2245-70.

[163] Watanabe, F., and Muguruma, H., "Ultimate Strength and Deformations of RC Panels," *Proceedings* of the sessions related to structural design, analysis and testing, ASCE Structures Congress 1989, pp. 31-38.

[164] Wernli, M., and Seible, F., "Assessment of Advanced Composite Stay Cable Systems," *Proceedings* of the Second International Conference on Composites in Infrastructure (ICCI '98), Vol. 1, Tucson, 1998, pp. 504-517.

[165] Westergaard, H.M., "On the Method of Complementary Energy and its Application to Structures Stressed Beyond the Proportional Limit, to Buckling and Vibrations, and to Suspension Bridges," *Proceedings* of the American Society of Civil Engineers, Vol. 68, No. 2, Feb. 1942, Transactions No. 107, pp. 765-793.

[166] Yagenji, A., "Experimental Study on Axial Compression Behavior of Circular RC Columns Using High Strength Reinforcement (in Japanese)," *Summaries of Technical Papers of Annual Meeting*, Architectural Institute of Japan, Structures II, Vol. C, 1992, pp. 547-548.

[167] Yamaguchi, T., and Naganuma, K., "Pure Shear Loading Tests of Reinforced Concrete Panels (in Japanese)," *Summaries of Technical Papers of Annual Meeting*, Architectural Institute of Japan, Structures II, Vol. C, 1988, pp. 545-548.

[168] Zhang, L.X., "Constitutive Laws of Reinforced Elements with Medium-High Strength Concrete," *M.Sc. Thesis*, Department of Civil and Environmental Engineering, University of Houston, Aug. 1992, 214 pp.

[169] Zhang, L.X., "Constitutive Laws of Reinforced Membrane Elements with High Strength Concrete," *Ph.D. Thesis*, Department of Civil and Environmental Engineering, University of Houston, Aug. 1995, 303 pp.

[170] Zhang, L.X., and Hsu, T.T.C., "Behavior and Analysis of 100 MPa Concrete Membrane Elements," *Journal of Structural Engineering*, ASCE, Vol. 124, No. 1, Jan. 1998, pp. 24-34.

[171] Ziegler, H., "On the Theory of the Plastic Potential," *Quarterly of Applied Mathematics*, Vol. 19, 1961, pp. 39-44.

Notation

Roman capital letters

A	area
D	dissipation
E	modulus of elasticity
F	actions in general, force
$\boldsymbol{F}$	vector of generalised forces
G	fracture energy
$\boldsymbol{K}$	stiffness matrix
M	bending moment
N	axial force
P	prestressing force
Q	pole of Mohr's circle
$\boldsymbol{S}$	total stress vector
U	(strain) energy
V	shear force
W	work
$\boldsymbol{Z}$	vector of generalised stresses

Roman lower case letters

a	distance
b	width, distance
c	cohesion, coefficient, distance
d	depth, distance
e	eccentricity, distance
f	material strength; distributed load
$\boldsymbol{f}$	vector of generalised deformations
g	function
h	height
k	coefficient
n	(normal) coordinate, number, modular ratio ($= E_s/E_c$)
p	$= (\sigma_1 + \sigma_3)/2$ (abscissa of the centre of the governing stress circle, $\sigma_1 \geq \sigma_2 \geq \sigma_3$); distributed load
q	$= (\sigma_1 - \sigma_3)/2$ (maximum shear stress, $\sigma_1 \geq \sigma_2 \geq \sigma_3$); distributed load
r	radius of curvature
s	spacing, distance
t	(tangential) coordinate, thickness
u, v	displacements
x	(axial/horizontal) coordinate
y	coordinate
z	(tangential/vertical) coordinate
$\boldsymbol{z}$	vector of generalised strains

Greek letters

α	angle
β	angle
γ	shear strain
Δ	difference, increment
δ	bond slip, displacement
$\boldsymbol{\delta}$	displacement vector
ε	axial strain
$\boldsymbol{\varepsilon}$	vector of strains
ζ	(relative) coordinate
θ	angle (between x- and 3-axes)
κ	non-negative coefficient
λ	coefficient
μ	coefficient of friction
ν	Poisson's ratio
ξ	(relative) coordinate
ρ	geometrical reinforcement ratio
σ	axial stress
$\boldsymbol{\sigma}$	vector of stresses
τ	shear stress
φ	angle of internal friction
Φ	yield function (plastic potential)
ω	mechanical reinforcement ratio $= \rho f_y / f_c$

Subscripts

b	bond
c	concrete, compression
d	dissipation
e	end of interval
F	actions in general, applied loads; fracture process
h	hardening
k	k-th element
m	mean value; mortar matrix
n	(normal) coordinate; nodal zone
o	initial, zero, reference value
p	prestressing steel
r	crack, cracking
s	reinforcing steel
t	(tangential) coordinate, tension
u	ultimate
v	shear, vertical
w	web
x	(axial/horizontal) coordinate
y	coordinate, yield
z	(tangential/vertical) coordinate
1,2,3	principal directions
(1)	first invariant
I,II	characteristic directions
av	average
$calc$	calculation
exp	experimental
inf	lower (chord)
max	maximum
min	minimum
nom	nominal
sup	upper (chord)

Superscripts

(c)	concrete
(p)	plastic
(r)	crack

Mathematical and special symbols

$\emptyset$	reinforcing bar diameter
,	partial derivative (with respect to the variable following the comma)
*	complementary; virtual cylinder, compression
˙	rate, velocity
~	representative value; effective (reinforcement ratio)

Berichte des IBK beim Birkhäuser Verlag Basel

Die aufgeführten Berichte sind unter Angabe der ISBN-Nr. direkt beim Birkhäuser Verlag Basel (Auslieferungsstelle) zu bestellen.
Adresse: Postfach 205, CH-4105 Biel-Benken (Tel. 061 721 77 84; Fax 061 721 79 50).

The listed publications can be ordered directly from Birkhäuser Publishers Basel by specifying the ISBN No.
Dispach address: Birkhäuser Verlag, P.O. Box 205, CH-4105 Biel-Benken/Switzerland
Tel. ++ 41 61 721 77 84; Fax. ++ 41 61 721 79 50

Manuel Alvarez, Peter Marti:
Versuche zum Verbundverhalten von Stahlbeton bei plastischen Verformungen
Bericht IBK Nr. 222, September 1996, ISBN 3-7643-5647-2, Fr. 78.--

Leena Eskola:
Zur Ermüdung teilweise vorgespannter Betontragwerke
Bericht IBK Nr. 223, September 1996, ISBN 3-7643-5653-7, Fr. 78.--

Tadeusz Szczesiak:
Die Komplementärmethode: Ein neues Verfahren in der dynamischen Boden-Struktur-Interaktion
Bericht IBK Nr. 224, September 1996, ISBN 3-7643-5655-3, Fr. 68.--

Seyed Mohammad Reza Tabatabai:
Finite Element-based Elasto-Plastic Optimum Reinforcement Dimensioning of Spatial Concrete Panel Structures
Bericht IBK Nr. 225, November 1996, ISBN 3-7643-5684-7, Fr. 78.--

Walter Kaufmann, Peter Marti:
Versuche an Stahlbetonträgern unter Normal- und Querkraft
Bericht IBK Nr. 226, November 1996, ISBN 3-7643-5687-1, Fr. 68.--

Philipp Stoffel, Peter Marti:
Modellversuche Europabrücke
Bericht IBK Nr. 227, März 1997, ISBN 3-7643-5762-2, Fr. 68.--

Marcus Schenkel, Thomas Vogel:
Versuche zum Verbundverhalten von Bewehrung bei mangelhafter Betondeckung
Bericht IBK Nr. 228, Mai 1997, ISBN 3-7643-5769-X, Fr. 58.--

Ralf Martens:
Zum Tragverhalten von Betonplatten mit integrierten Schalungselementen
Bericht IBK Nr. 229, Juni 1997, ISBN 3-7643-5797-5, Fr. 108.–

Matthias Barth, Peter Marti:
Versuche an knirsch vermauertem Backsteinmauerwerk
Bericht IBK Nr. 230, August 1997, ISBN 3-7643-5812-2, Fr. 48.–

Stefan Köppel, Thomas Vogel:
Feldversuch Steilerbachbrücke
Bericht IBK Nr. 231, September 1997, ISBN 3-7643-5838-6, Fr. 48.--

Glauco Feltrin:
Absorbing Boundaries for the Time-Domain Analysis of Dam-Reservoir-Foundation Systems
Bericht IBK Nr. 232, November 1997, ISBN 3-7643-5869-6, Fr. 108.--

Walter Borgogno:
Tragverhalten von Slim Floor Decken mit Betonhohlplatten bei Raumtemperatur und Brandeinwirkungen
Bericht IBK Nr. 233, Dezember 1997, ISBN 3-7643-5899-8, Fr.68.--